SpringerBriefs in Business

SpringerBriefs present concise summaries of cutting-edge research and practical applications across a wide spectrum of fields. Featuring compact volumes of 50 to 125 pages, the series covers a range of content from professional to academic. Typical topics might include:

- A timely report of state-of-the art analytical techniques
- A bridge between new research results, as published in journal articles, and a contextual literature review
- A snapshot of a hot or emerging topic
- An in-depth case study or clinical example
- A presentation of core concepts that students must understand in order to make independent contributions

SpringerBriefs in Business showcase emerging theory, empirical research, and practical application in management, finance, entrepreneurship, marketing, operations research, and related fields, from a global author community.

Briefs are characterized by fast, global electronic dissemination, standard publishing contracts, standardized manuscript preparation and formatting guidelines, and expedited production schedules.

Antonia Caro-Gonzalez

Transformative Governance for the Future

Navigating Profound Transitions

 Springer

Antonia Caro-Gonzalez
Eoh-for-Good
Bilbao, Spain

ISSN 2191-5482 ISSN 2191-5490 (electronic)
SpringerBriefs in Business
ISBN 978-3-031-43131-9 ISBN 978-3-031-43132-6 (eBook)
https://doi.org/10.1007/978-3-031-43132-6

© The Author(s) 2024. This book is an open access publication.

Open Access This book is licensed under the terms of the Creative Commons Attribution 4.0 International License (http://creativecommons.org/licenses/by/4.0/), which permits use, sharing, adaptation, distribution and reproduction in any medium or format, as long as you give appropriate credit to the original author(s) and the source, provide a link to the Creative Commons license and indicate if changes were made.
The images or other third party material in this book are included in the book's Creative Commons license, unless indicated otherwise in a credit line to the material. If material is not included in the book's Creative Commons license and your intended use is not permitted by statutory regulation or exceeds the permitted use, you will need to obtain permission directly from the copyright holder.
The use of general descriptive names, registered names, trademarks, service marks, etc. in this publication does not imply, even in the absence of a specific statement, that such names are exempt from the relevant protective laws and regulations and therefore free for general use.
The publisher, the authors, and the editors are safe to assume that the advice and information in this book are believed to be true and accurate at the date of publication. Neither the publisher nor the authors or the editors give a warranty, expressed or implied, with respect to the material contained herein or for any errors or omissions that may have been made. The publisher remains neutral with regard to jurisdictional claims in published maps and institutional affiliations.

This Springer imprint is published by the registered company Springer Nature Switzerland AG
The registered company address is: Gewerbestrasse 11, 6330 Cham, Switzerland

Paper in this product is recyclable.

Preface

Humanity is in transition. Our organisations, our innovation ecosystems are in transition. The whole planet is in transition. Welcome to our Anthropocene era.

But the truth is that in general humans don't like changes. We see the clouds, sometimes very black ones, but it is not usually until the storm bursts heavily on us that we react and run to find a safe shelter.

We face continuous processes of change at different levels: a change of position within the company, a family sickness, the deployment of a new digital system. We could give hundreds of examples of small or bigger switches that alter our lives. In general, fear prevents us from being brave and acting on time. We prefer to stay in the comfort zone to confront the things that do not work, or we perceive that are evolving in the wrong direction.

On a wider scope, we've recent examples of these 'heavy storms' that threaten our established socio, economic, political, health, educational, financial status quo: a recent world-wide pandemic, climate change with more frequent droughts, heavy floods, or wildfires; water scarcity; plastic pollution; wider social unbalances with two deep financial crises in less than 20 years. We find specific intersections of time and space in which many lives, companies, projects, ecosystems are at stake.

But the new thing is that all that these changes are anthropogenic in great measure. We are generating them.

Carl Sagan at the end of his marvellous TV series Cosmos asked: Who speaks in the name of the planet?

Now we need to have the **courage to recognise that we, humans, are the agents of such changes, for bad or for good. Our choice is for good, for a new common good. We need shared visions about this goal**; we need new generations of social, digital and environmental innovators and **entrepreneurs**, committed to boost collaborations; we need to create **new transformative governance, based in new** methodologies, **tools and services** to respond to these changes. Briefly, we urgently need a global and personal wake up to generate real transformations of our societies, our artificial systems and our natural ecosystems for which we are responsible.

The trigger is: if we can envision positive changes and win-win situations, let's try! Let's try together by adding our 'one drop in the ocean' to a co-dreamt better world.

Bilbao, Spain   Antonia Caro-Gonzalez

Acknowledgements

To the many, many international colleagues who have been part of this journey, you are not just collaborators but close friends and kindred spirits. Your unwavering support, invaluable insights and dedication have enriched this work beyond measure. Special thanks goes to Jorge Iván Contreras Cardeño, Renata Petrevska-Neschoska, Artur Serra and José Sierra Villarejo, whose contributions and reviews have significantly shaped the content and direction of this book.

A sincere acknowledgement goes to Analogias, and specifically to Sofi, for her exceptional design expertise that has breathed life into the visual representation of my ideas. Your creativity and artistic flair have elevated the impact of this work.

To my family, whose constant love, encouragement and understanding have been a pillar of strength throughout this endeavour, I am deeply grateful. Your unwavering belief in me has been the driving force behind this project, and I am forever indebted to each of you for your support.

I extend my deep appreciation to the Eoh-for-Good team, whose collaborative spirit and collective efforts have contributed to the success of this project. Your passion for transformative governance and commitment to creating positive change is a true inspiration.

I want to extend my heartfelt gratitude to all the scholars, researchers and practitioners whose works have informed this exploration. May our collective efforts contribute to a brighter future, where transformative governance guides us towards a world that thrives on justice, sustainability and innovation.

Lastly, I would like to express my gratitude to all those whose names may not appear on these pages but have provided invaluable support, feedback and encouragement. Your contributions, whether big or small, have played an essential role in shaping this book.

This work would not have been possible without the collective effort, generosity and enthusiasm of everyone mentioned above and countless others who have played a role in this journey. Your involvement has made this book a truly collaborative endeavour, reflecting the power of unity and shared purpose.

As we navigate the profound transitions ahead, may our collective commitment to transformative governance continue to foster positive change in the world. Together, we can build a more just, sustainable and inclusive future for all.

Finally, I am deeply honoured to extend my heartfelt gratitude to the three organisations that have demonstrated a remarkable commitment to advancing knowledge and fostering innovation in their respective domains, and I am privileged to have the opportunity to collaborate with: Valencian International University, the Digital Society Technology Unit at the I2Cat Foundation and MAVERICKS Advancing Higher Education.

Toñi Caro

A Rapid Rundown on Crucial Terms

Anthropogenic
Anthropogenic refers to the significant and dominant human influence on the environment, contrasting the notion of an independent and timeless nature. In recent years, scientific evidence has shown that human activities have become the primary driving force behind global and geological changes, leading to the recognition of a new geological epoch called the Anthropocene. Anthropogenic global warming is a central concern in this context, prompting international efforts to prevent detrimental interference with the climate system. Over time, anthropological research has been crucial in understanding the complex interplay between humans and the environment, encompassing various scales from the molecular to the global, and contributing to transdisciplinary efforts in this area [1].

Col·Laboratory[1]
It is a neologism to express the next generation of quadruple or n-helix living labs. They are cooperative instruments based on a peer-to-peer approach between business/social/technology-driven innovations.

Difference Between Social and Societal Impact
The difference between social and societal impact lies in the scope and scale of their effects. Social impact refers to the specific and direct outcomes or changes resulting from a particular intervention, programme or initiative. It focuses on addressing immediate social issues and improving the well-being of individuals or specific communities. Social impact is often localised and can be measured through metrics such as improved access to education, healthcare or reduced poverty levels.

On the other hand, societal impact encompasses a broader and more systemic perspective. It pertains to the broader changes and transformations that influence society as a whole. Societal impact involves addressing larger-scale challenges and tackling complex issues that affect multiple aspects of human life, such as systemic inequality, climate change or technological disruptions. Measuring societal impact may require considering a wide range of interconnected factors and long-term effects on various stakeholders.

---

[1] https://integercollab.eu/what-is-integer/, accessed June 13, 2023.

Dynamic Capability

It is the firm's ability to integrate, build and reconfigure internal and external competences to address rapidly changing environments.[2]

Quadruple Helix

The quadruple helix refers to a collaborative innovation model that involves four key stakeholders: academia, industry, government and civil society. This framework aims to foster innovation and address complex challenges by promoting active engagement and interaction among these actors. In addition to the traditional triple helix model (university-industry-government), the inclusion of civil society adds a crucial dimension, ensuring that societal needs and perspectives are integrated into the innovation process. The quadruple helix model aims to create more dynamic and inclusive ecosystems that facilitate knowledge exchange, co-creation and the development of sustainable solutions for the benefit of society.[3]

Social-Digital Innovation

According to the European Commission,[4] social innovation 'aims to advance European life through improving working conditions, education, community development or health, or through tackling critical problems such as poverty or discrimination'.

Social-digital innovation refers to the creative and transformative use of digital technologies and social practices to address societal challenges and enhance the well-being of individuals and communities. It involves leveraging digital tools, such as social media, data analytics, artificial intelligence, and emerging technologies, to develop novel solutions for social issues like healthcare, education, poverty and inclusivity. Through the combination of technology and a deep understanding of human needs and social dynamics, social-digital innovation promotes positive social change, connectivity and collaboration, paving the way for a more inclusive, equitable and sustainable future.

References

1. European Commission. (2023) COMMUNICATION FROM THE COMMISSION TO THE EUROPEAN PARLIAMENT AND THE COUNCIL: Sustainability and people's wellbeing at the heart of Europe's Open Strategic Autonomy (No. COM(2023) 376 final) (p. 21). Brussels. https://doi.org/10.1163/2210-7975_HRD-4679-0058
2. Mazzucato M (2018) Mission-oriented innovation policies: challenges and opportunities 27(5):803–815. https://doi.org/10.1093/icc/dty034

---

[2] https://www.davidjteece.com/dynamic-capabilities, accessed May 30, 2023.

[3] Mazzucato's emphasis on purposeful innovation, public–private partnerships and value creation challenges entrepreneurs to consider the broader societal implications of their ventures. By actively collaborating with diverse constituents, entrepreneurs can contribute to sustainable economic growth, social equity and environmental stewardship [2]

[4] https://ec.europa.eu/european-social-fund-plus/en/social-innovation-and-transnational-cooperation#:~:text=Social%20innovation%20aims%20to%20advance,such%20as%20poverty%20or%20discrimination, accessed June 13th, 2023.

Contents

1 Introduction 1
2 A Call for Action—Tackling the Profound Transitions Ahead 7
   2.1 The Urgency to Plan Ahead the Transition Journey 8
   2.2 Towards a Just Triple Transition: Social, Green and Digital 16
   References 21
3 Tailoring Transformative Governance for the Common Good 23
   3.1 A Compass in a Nutshell—The Navigating Tool 23
   3.2 Grounding Approaches and Principles for a Just Triple
        Transition 24
        3.2.1 The Travel Arrow Pointing Towards the Common
              Good 28
        3.2.2 Approaches 31
        3.2.3 Principles 36
   3.3 Long-Term Goal-Aligned Alternatives 40
   References 44
4 Establishing a Culture of Innovation and Risk-Taking 47
   4.1 Fostering Intra-entrepreneurship for Individuals to Become
        Agents of Change 49
   4.2 The Role of Leadership in Cultivating Intra-entrepreneurship 50
   4.3 Creating Intra and Entrepreneurial Just and Inclusive
        Dynamics 52
   References 55
5 Driving Systemic and Multi-level Transformative Governance 57
   5.1 An Overview of the Multi-i Collaborations
        for Transition—20+ Dimensions that Start with An 'i' 58
   5.2 The Deployment of Multi-level Innovative Governance
        Dynamics—The 10 Collaborative Elements that Start
        with An 'i' 60
   5.3 Forming Multi-i Collaborations 73

		5.3.1	How to Start Processes of Change	75
		5.3.2	Navigating Multi-level and Multi-actor Innovative Governance Processes	77
		5.3.3	Maximising the Learning and Innovation that Occurs in Multi-level and Multi-agent Collaborative Governance Processes	82
	5.4		Reinforcing the Transition Gap—From Established to Emerging Systems	83
	References			88
6	**Designing Ad Hoc Impact Monitoring Systems**			91
	6.1		The Importance of Measuring Progress and Tracking Impact Across Dimensions	94
	6.2		How Can We Design Impact Follow-Up Systems and an Ad Hoc Battery of Indicators?	96
	References			100
7	**Conclusion: Organisations and Ecosystems in Transition—Nurturing Transformative Governance**			103
Bibliography				107

Abbreviations

CoR	European Committee of the Regions
DNA	Deoxyribonucleic Acid
EC	European Commission
ENIL	European Network of Intergenerational Learning
ERA	European Research Area
ESG	Environmental, Social and Governance
EU	European Union
GDP	Gross Domestic Product
HCD	Human-Centred Design
IFRS	International Financial Reporting Standards
KPI	Key Performance Indicators
LERU	League of European Research Universities
PhD	Philosophie Doctor
R&I	Research and Innovation
RTOs	Recovery Time Objective
SASB	Sustainability Accounting Standards Board
SDGs	Sustainable Development Goals
SMEs	Small and Medium-sized Enterprises
UN	United Nations

List of Figures

Fig. 2.1	Picture in Bolzano	8
Fig. 2.2	Transition pathways	9
Fig. 2.3	Reinforce bridge between established and emerging systems	11
Fig. 2.4	Key components of innovation ecosystems	13
Fig. 2.5	Four (or n-) helix collaboration bridge	15
Fig. 2.6	Priority areas of action	18
Fig. 2.7	Twinning the green and digital transitions	19
Fig. 2.8	MoskEUteers for a just triple transition	20
Fig. 3.1	Eoh-for-good compass for social transformation	25
Fig. 3.2	Current policies and transition management	42
Fig. 3.3	Amartya Sen, Martha Nussbaum and Henry Richardson, hosting of the conference of the human development and capability association (HDCA) held in September 2015	43
Fig. 4.1	Full moon night at Serignan Plage (own picture)	51
Fig. 5.1	Overview of the 20+ 'i' dimensions	59
Fig. 5.2	The 10 collaborative 'i's	61
Fig. 5.3	Multi-i co-creative governance tornado	74
Fig. 5.4	Stages of the rotating movement	75
Fig. 5.5	Stages of the rotating movement and particles	76
Fig. 5.6	Ten collaborative 'i's with the movement of co-operations	78
Fig. 5.7	Multi-'i' co-creative governance tornado	79
Fig. 5.8	Multi-'i' co-creative governance tornado and concrete initiatives	80
Fig. 5.9	Three different pathways from collaborative dynamics	82
Fig. 5.10	Overview of the process of transition © Eoh-for-good	84
Fig. 5.11	Snapshot of the emerging phase © Eoh-for-good	85
Fig. 5.12	Ten collaborative 'i's © Eoh-for-good	85
Fig. 5.13	Snapshot of the spanning power of multi-i co-creative vortices © Eoh-for-good	86
Fig. 5.14	Mediators and balancers	87
Fig. 6.1	ESG graph © Eoh-for-good	92

List of Tables

Table 3.1	Eoh-for-good compass features	26
Table 3.2	Elements 1, 2 and 3 of the compass	29
Table 3.3	Element 4 of the compass	32
Table 3.4	Element 5 of the compass	40
Table 3.5	Element 6 of the compass	41
Table 5.1	Element 7 of the compass	77
Table 5.2	Element 8 of the compass	78
Table 6.1	Element 10 of the compass	95
Table 6.2	Elements 9 and 11 of the compass	98

Chapter 1
Introduction

Welcome to the book 'Transformative Governance for the Future: Navigating Profound Transitions'. In this book, we embark on an exploration of the critical aspects of managing change and navigating the complexities of emerging systems and innovations. The challenges facing humanity and the planet demand a radical shift in our approach—one that goes beyond traditional boundaries and embraces a holistic perspective to intentionally lead towards a desired state or outcome, considering the complexities, challenges and opportunities that arise along the way. Eoh-for-Good—Leading Systemic Transformation for the Common Good is the result of many years of dreaming, experimenting and implementing systemic ways to drive transformations for the common good.

This is envisaged to share some of these experiences, innovative practices and lessons learned along amazing individual, but mostly collective,[1] journeys of experimentation. All this while navigating change at different levels: personal, group, institutional and ecosystem, or at the juncture between them.

Everything started with an attitude to constantly improve. Motivation, or better to say, a constant search for never losing motivation, has driven my personal and professional life. An entrepreneurial spirit and innate orientation to continuous learning have engined the way. Early in life, learning, and especially learning by doing with others, became my passion (we built authentic farms, castles, cities… with Exin Castillos,[2] Playmobil and Lego bricks with my sister and brothers). We learned to imagine, plan, negotiate, build, de-build and rebuild again. Playing gave us the ground to learn many skills that have proven crucial later in our personal and professional careers. I have not stopped learning and hope not to do so. I love finding and proposing new things and ways forward.

Later, in my professional life, proposing new ways to overcome difficulties, positiveness and a committed team spirit became the DNA of our team. We addressed the challenges we faced in our everyday work, by putting together our capacities,

---

[1] This is the reason for changing from the I to the we during the explanation in the book.

[2] Exin Castillos is a Spanish construction game created in the late 1960s (1968).

© The Author(s) 2024
A. Caro-Gonzalez, *Transformative Governance for the Future*,
SpringerBriefs in Business, https://doi.org/10.1007/978-3-031-43132-6_1

diverse profiles, personalities, creativity and different approaches to doing things. The quest for complementarities and caring criticism help us grow as professionals and as an energetic team. This way, our regular duties: search for funding, improve management systems to become more efficient, encompass a) different rhythms, b) egos in an academic setting, c) interests in the wider innovation ecosystem, … etc. These became a trigger that helped us overcome the burden of long working hours, unleashing our creativity. Ultimately, solving these challenges and tensions got ourselves at the service of change.

Our main mission was to support researchers in getting international funding, managing projects… For doing so, we needed to address other issues alongside: how to understand and translate policy priorities into actionable research and innovation; how to envisage and invent new ways to bring research results closer to society and to policy decision-making and how to show and demonstrate the impact of research and innovation (R&I).

> This entails having the vision and moving from intuited ideas to actions, from dreams to initiatives that get changes in motion. Trial and error and testing make progress possible until things get consolidated, 'institutionalised' or appropriated.

Having combined learning by doing with a conscious reflection on our everyday performance (what things work, which didn't, what if we try a new thing? has carved my passion to improve, to learn, to learn by experimenting and doing. What a passion ideates, design, innovate, reflect and implement on and in action!

The book offers the lessons learned during the last 20 years to help leaders in different positions to reflect on their challenges and to draw flexible governing roadmaps within their organisation, division, unit, department, ecosystem…

It proposes thoughtful inquiries fostering a proactive and engaged approach to problem-solving and decision-making. This way managers, leaders and entrepreneurs are triggered to propose paths for organisations and/or ecosystems to find their way to effective and efficient execution of innovative solutions rather than relying on generic cut-and-paste responses. The following questions have triggered and imprinted my personal and collective experimentation over the years. However, each agent of change needs to make their own questions, adapted to their own settings, aims, contexts, vision, etc.

- What role can my organisation, our board of directors, department, research unit, the sector I work in, the ecosystem I'm part of play?
- What if I (each of us) embrace responsible and responsive performance well in advance?
- And, linked with the purpose of this book, how can we navigate the transitions? the changes we face? and what if we do it by pursuing the common good?

1 Introduction

These can be just some questions for us to think about our personal positioning, our capacity to respond, our own responsibility from our role, from our capabilities, agency and commitment.

Eoh-for-Good name is made up of three parts,[3] which are our DNA:

- **eo**: From Latin: it denotes movement - advance. It resembles a wake call to all of us eoooo, wake up!!
- **h**: It is a representation of humanity. People at the centre, as a fundamental part of our processes.
- **For good**: The common good is not an option. It is our main objective; it is in our DNA. For good means two things, for the good and forever, to sustain it perpetually with the necessary adjustments as we move along.

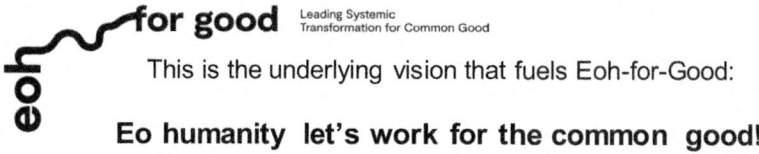

This is the underlying vision that fuels Eoh-for-Good:

Eo humanity let's work for the common good!

As Fran Albalá, my dear husband, explains in his talks, we need to be ADI,[4] be alert and attentive to what is happening around us, to the way reality is constantly changing the rules of the game, facing us with new challenges. This forces us to adapt the way we need to respond and improve the situations.

We need more fluid and agile management that leverages intra-entrepreneurship, promotes synergies, capitalises diverse perspectives and focuses on the common good. This is becoming critical as we move towards:

- More conscious and demanding citizens with business and production practices,
- More demanding regulations aligned with 'not harm' principles,
- Better and more refined accountability systems on the return on investment, etc.

As the world faces pressing challenges and transitions on social, environmental and technological fronts, it becomes increasingly essential to plan ahead and equip us with the necessary tools and knowledge to steer these transformations towards a more just and sustainable future.

The book is structured in five chapters:

In the Chap. 2, we articulate the urgency of acting and preparing for the journey of transition. We delve into the factors that reinforce the transition gap, bridging the chasm between established and emerging systems and innovations. An intriguing highlight of this chapter is the exploration of the multidimensional co-creation vortices of transition—a collection of 20+ dimensions that all start with an 'i,' shedding light on the diverse facets of transformative change.

---

[3] Heartfelt gratitude to Ouidesign for their exceptional professionalism in giving life to our brand.
[4] Adi is the word in Basque that means 'pay attention'.

We stress the significance of foresight and proactive planning to address the challenges that lie ahead. By understanding the importance of tackling transitions head-on, we set the stage for the subsequent discussions on transformative governance. The chapter concludes with an essential discussion on the concept of a just triple transition, underlying the need to encompass the social, environmental and technological dimensions. This approach highlights the interconnectedness of these three pillars, setting the stage for inclusive and sustainable transformations.

In Chap. 3, we present a compass in a nutshell—a navigating tool for transformative governance. This compass guides us through grounding approaches and principles that steer the transition towards the common good. By understanding the travel arrow pointing towards the common good, we explore transformative governance from an inclusive and sustainable perspective. We seek to equip leaders as agents of change with the knowledge to navigate the complexities of transformative change, fostering just and equitable outcomes for society.

Chapter 4 centres on encouraging a culture of innovation and risk-taking as crucial elements of transformative governance. We delve into the concept of intra-entrepreneurship, entrusting individuals to become agents of change within their organisations and communities. Additionally, we explore the pivotal role of leadership in cultivating an environment that encourages and supports innovation. By creating just and inclusive dynamics, we aim to lay the foundation for a transformative and adaptive organisational culture.

In Chap. 5, we delve into the driving forces behind systemic transformative governance, focusing on the collaborative elements denoted by the mentioned dimensions that start with an 'i'. These elements form the foundation of co-creative vortices that promote collective innovation and change on multiple levels. They are the building blocks for effective multi-actor collaborations and partnerships. We explore how these vortices of collaboration can maximise the impact of governance processes and navigate innovation effectively, enabling us to address complex challenges and drive transformative change.

Chapter 6 explores the significance of evidence-based decision-making, providing valuable insights into creating comprehensive impact monitoring systems and a battery of indicators to assess the outcomes of transformative governance efforts. We highlight the importance of following up progress and tracking the impact of transformative governance initiatives across various dimensions. When agents committed to change invest efforts and resources in designing ad hoc impact monitoring systems, they can ensure that their efforts are better aligned with the overarching goals of a just, sustainable and equitable future.

Open Access This chapter is licensed under the terms of the Creative Commons Attribution 4.0 International License (http://creativecommons.org/licenses/by/4.0/), which permits use, sharing, adaptation, distribution and reproduction in any medium or format, as long as you give appropriate credit to the original author(s) and the source, provide a link to the Creative Commons license and indicate if changes were made.

The images or other third party material in this chapter are included in the chapter's Creative Commons license, unless indicated otherwise in a credit line to the material. If material is not included in the chapter's Creative Commons license and your intended use is not permitted by statutory regulation or exceeds the permitted use, you will need to obtain permission directly from the copyright holder.

Chapter 2
A Call for Action—Tackling the Profound Transitions Ahead

> *Those who feel they have reached the pinnacle, and no longer acquire sufficient knowledge, have only a remaining option: to descend.*

In this chapter, we embark on a compelling journey, exploring the urgency of acting and preparing ourselves for the deep transitions we are immersed in.

We underline the significance of foresight and proactive planning to navigate the challenges that await us. Understanding the importance of confronting transitions head-on lays the foundation for the subsequent discussions on transformative governance.

This chapter engages in a crucial dialogue on the concept of a just triple transition that focuses on the need to balance social, environmental and digital dimensions. Recognising the interconnection between these sets the stage for inclusive and sustainable transformations that will shape our collective future.

Let us embark on this enlightening journey together, as we prepare ourselves for the profound transitions that lie ahead!

Individuals and institutions have limitations in addressing complex challenges, therefore fostering change and transformative initiatives towards a just triple transition is crucial for achieving shared outcomes and impacts. This involves (1) recognising the need for improvement, (2) driving institutional change with a systemic approach, (3) rethinking governance structures, (4) promoting transparency and accountability and (5) promoting a culture of collaboration and shared responsibility.

Actors committed to change can identify areas where institutional advancements are required.

2.1 The Urgency to Plan Ahead the Transition Journey

Innovation and change are like climbing a mountain. We can stay down or try to conquer the top, even at the risk of falling. Perhaps at the top, we will find virgin and unexplored and better spaces for fairer human development, more flourishing market opportunities, regenerative outcomes for the planet, etc. (Fig. 2.1).

However, we need to become aware, to overcome fear to face the head-on challenges, knowing that these will make us emerge stronger and better than we were. We can embrace life's storms as opportunities to soar higher.

Traditional linear approaches to innovation offer a straightforward and predictable path from idea to execution. While they have their strengths, they often fall short in addressing the intricate nature, uncertainties and interconnections of the many opportunities and challenges of society, technology, economy and the environment. Linear thinking can impede the capacity of individuals and organisations to adapt, collaborate and effectively respond to evolving needs with a challenge-driven mindset.

As portrayed in the illustration, few embrace the nonlinear, complex and demanding path. Addressing the sloppy path may help organisations and innovation ecosystems, and each of us, to achieve milestones towards new goals. The complex, sloppier and more bending path that gets us out of our comfort zone and puts in motion efforts to address the deep transitions we are involved in (Fig. 2.2).

But what and how is this journey we are talking about? This is the one that individuals, companies, entities and ecosystems need to plan well to address the changes ahead. As in the figure portrayed above, the transition is a steep road

Fig. 2.1 Picture in Bolzano

2.1 The Urgency to Plan Ahead the Transition Journey

Fig. 2.2 Transition pathways

full of twists and turns, ups and downs or green and red lines. In the processes of change, we find two realities that co-exist in constant evolution and adjustment with opportunities and tensions between established and emerging systems.

Throughout this process, it is important to anticipate recessions and be prepared to adapt and switch roles. This flexibility allows organisations and ecosystems to navigate challenges and seize opportunities for growth and development.

Fear, resistances, lack of coordination and inertias are just a few features that restrain us to think out of the box and chose the slopy but right complex path.

Instead of fixating solely on sales volume, a company must prioritise safeguarding its margins.

Choosing the complex, but just transition path, helps us shift from established systems in organisations and ecosystems towards the very much-needed renewed emerging systems.

Deteriorating margins push us perilously closer to the edge. The purpose is that we get better equipped to tackle the challenges and find innovative ways and solutions in an ever-changing world. To maintain healthy margins, it is crucial to position our products, services or policies as more than commodities and demonstrate clear differentiation.

The success of a company or any endeavour hinges on generating value for customers or users and ensuring that a portion of that value is retained. Let's embark on your transformational journey towards sustainable success.

No matter which role, which type of company, which position we play in a public service or in the private sector, in society (mother, father, child, young, adult, active worker, retired), how could we derive fulfilment from confronting challenges and transforming them into steppingstones towards achievement and success?

This calls for adopting a more flexible and comprehensive mindset that requires a more holistic approach encompassing short-, medium- and long-term innovations.

Eoh-for-Good's flexible and adaptable methodology can align individual and institutional needs, interests and innovations into multi-i co-creative governance vortices to transform challenges into solutions through multi-actors' collaborative endeavours.

At Eoh-for-Good, we combine over 20 years of experience managing R&I internationalisation, outreach, engagement and impact strategy with a collaborative approach involving 1300 stakeholders across various sectors. We offer actionable areas for action, including boosting the global presence and sphere of influence of EU organisations and companies.

Institutional and ecosystem change involves reimagining and transforming institutions to be more inclusive, accountable and responsive to the needs of society and the environment. This may involve changes in decision-making processes, resource allocation and organisational culture to prioritise fair sustainable development. It entails the alignment of policies, regulations and governance structures with the common good goals.

Aligning innovative institutional processes of transformation towards a courageous goal-alignment for the synchronisation and maximisation of impact. Enables actors committed to change to focus their resources and initiatives on areas where they can make the most significant contributions and helps create a coherent and synergistic approach, where innovation drives institutional change and institutional change enables an environment for innovation.

> Contrary to the traditional notion that cooperation, sustainability and competitiveness are mutually exclusive, encouraging sustainable collaborative just practices can enhance the attractiveness and competitiveness of organisations and innovation ecosystems. Actors can support each other and learn from each other's experiences. Fairer sustainable innovation can lead to cost savings, increased efficiency and improved resource management, providing a competitive advantage to companies and start-ups.

By integrating just collaborative sustainability into operations, products and services, regions, companies, clusters, entrepreneurs can differentiate in the market, attract environmentally conscious consumers and better respond to evolving regulatory requirements (e.g. EU Directive on reduction of the impact of certain plastic products on the environment).

Sustainable-just-oriented innovation can also open new market opportunities, particularly in sectors focused on health and well-being, renewable energy, clean technologies and sustainable solutions.

Figure 2.3 presents an overview of the transition path. Gaps arise between two systems in transition and the bridge between them is usually very weak and fails in establishing the necessary connections.

2.1 The Urgency to Plan Ahead the Transition Journey

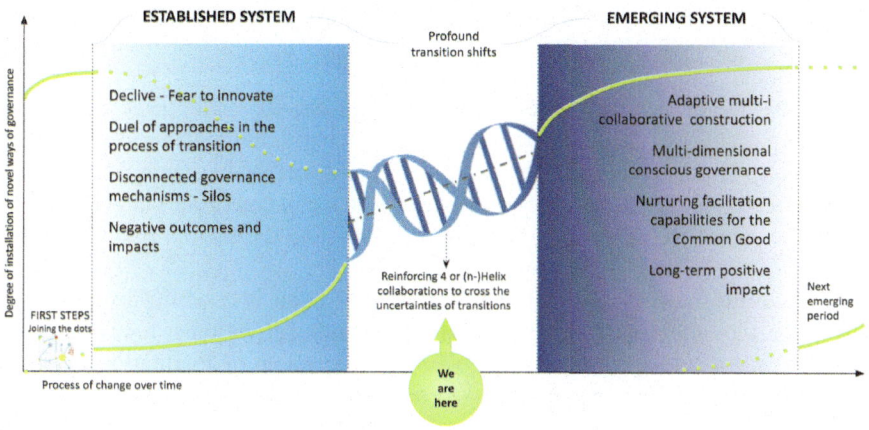

Fig. 2.3 Reinforce bridge between established and emerging systems

When reflecting on the processes needed for transformative governance in our era of transitions, we see this as profound transition shifts between established, when old ways of doing things are threatened, and emerging systems. To truly transform the economy, social relationships and values of society, we must navigate the profound transition shifts of our era towards the emerging systems. Novel governance mechanisms need to be employed, tested, refined to overcome the deep gap between the declining established systems and the urgent emerging ones. Many innovations often end up into deep gaps during the periods of transition or change.

It is evident that the current mechanisms for transitioning between systems are insufficient, leading to missed opportunities for innovation adoption which calls for a shift in our approach to transitions. As a result, many of our innovations, products and services fail to reach the market or society (dropping to the valley of death[1]). This phenomenon has been labelled as the 'European paradox of innovation'. The European paradox of innovation refers to the region's strong emphasis on research and development, but its struggle to effectively translate these innovations into successful commercial products or services, leading to a weakness in overall innovation competitiveness. To address this challenge, we require more robust transition and governance models that promote collaborations, strengthening the ups and downs of transitioning to emerging systems. There is a need for cohesive governance, collaboration and effective strategies that bridge this gap and ensure the successful integration of new ideas and technologies into the market and society.

---

[1] The 'valley of death' in innovation refers to the challenging phase where promising ideas or technologies struggle to secure funding or commercial viability before reaching the market.

Carlota [2], as expert in technology and socio-economic development, suggests that as new technological paradigms emerge, they bring about disruptive changes that challenge existing norms and practices. Perez portrays the breaks as turning points, due to the challenges associated with societal acceptance of changes and new paradigms.[2] This period is marked by what Schumpeter called 'creative destruction'. However, Perez emphasises that this is just the beginning.

The transition gap, bridging established systems and emerging innovations, is a pressing concern we delve into, seeking to understand the factors reinforcing this divide.

Successfully transitioning through this turning point is essential to realise the benefits of the new paradigm. To address this challenge, we must focus on developing transition and governance models that are capable of facilitating, as much as possible, smooth and seamless transitions. These models should encourage collaborations among various stakeholders, including industry leaders, policy makers, researchers and entrepreneurs.

When a diversity of actors creates and diffuses new ideas, technologies and practices, they are better prepared to prioritise the triple bottom line, considering economic, social and environmental dimensions and integrate sustainable practices into their operations and business models. They encourage collaboration, knowledge exchange and social cohesion, leading to impactful and socially just outcomes.

Through collective action, motivated innovators from different contexts, disciplines, expertise and sectors can challenge existing norms and drive systemic change.

Embracing shared responsibility to achieve the common good is essential at every stage of innovation, entrepreneurship and institutional change. This involves ensuring that socio-digital and technological advancements:

1. create societal value, considering the social and environmental impacts of products and services,
2. do not harm individuals, communities or the planet promoting fair trade practices,
3. contribute to human and environmental well-being and sustainability and
4. embrace entrepreneurship necessary for institutional change, prioritising transparency, accountability and the common good.

Natural and innovation ecosystems, and organisations lack a more encompassed way of various actors working together, including entrepreneurs, start-ups, citizens, established companies, research institutions, government agencies and supportive infrastructures. Innovation ecosystems are characterised by complex interdependencies and interactions among their components. The success and effectiveness of the ecosystem rely on the synergistic collaboration and exchange of knowledge, resources, policy areas, infrastructures and expertise among its

---

[2] WE have already discussed this issue in a paper in the press [3].

2.1 The Urgency to Plan Ahead the Transition Journey

actors. Organisations, companies and innovation ecosystems are dynamic and interconnected networks that need to better collaborate and interact to raise innovation and drive economic growth.

Inclusive innovative institutions and ecosystems break down barriers, provide equal opportunities and entrust individuals from all backgrounds to participate and contribute.

The collective commitment towards the common good and shared outcomes cultivates a more sustainable, competitive and inclusive future for all. Key components of innovation ecosystems include a wide variety of actors and interdependencies (Fig. 2.4):

1. **Knowledge and Research Institutions**

    Universities, research centres and academic institutions contribute to the creation and dissemination of knowledge, fostering research and development activities that drive innovation.

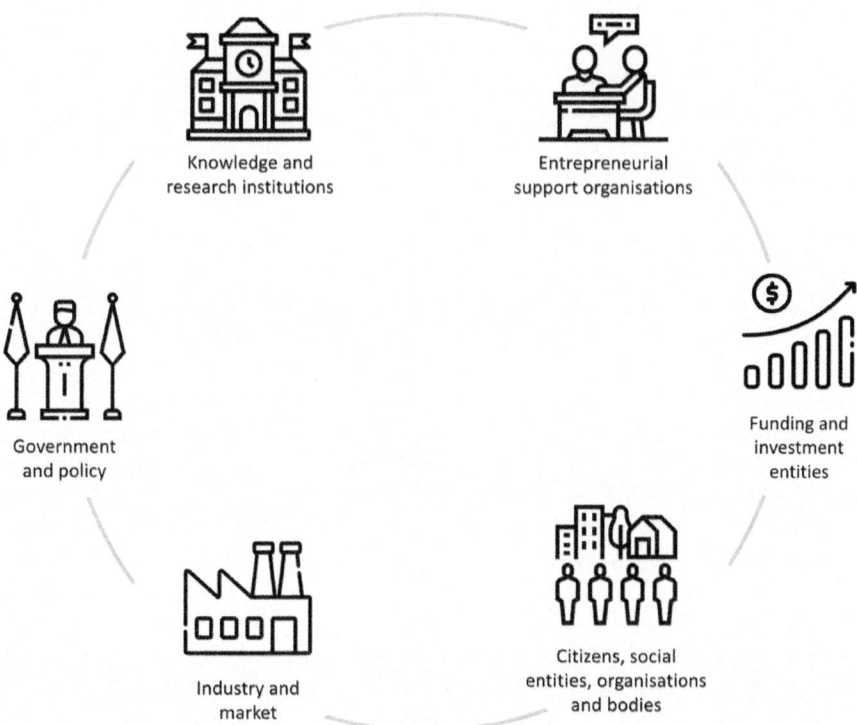

Fig. 2.4 Key components of innovation ecosystems

2. **Entrepreneurial Support Organisations**

 Incubators, accelerators and entrepreneurship centres provide resources, mentorship and guidance to start-ups and entrepreneurs, facilitating their growth and success.

3. **Funding and Investment**

 Access to capital and funding sources, such as venture capital firms, angel investors and government grants, is crucial for supporting innovation and scaling up entrepreneurial ventures.

4. **Industry and Market**

 Established companies, industries and markets provide opportunities for collaboration, commercialisation and the adoption of innovative solutions, contributing to the growth and sustainability of the ecosystem.

5. **Government and Policy**

 Government agencies play a role in shaping the regulatory environment, providing incentives and nurturing an ecosystem that supports innovation and entrepreneurship.

6. **Citizens, Social Entities, Organisations and Bodies**

 They play a pivotal role in processes of change by actively engaging, advocating, and collaborating. They contribute ideas, resources and collective action to shape and drive transformative initiatives, encouraging inclusivity, participation and a sense of ownership for positive societal impact.

The quadruple (or n-) helix approach advocates for a collaborative and participatory problem-solving approach, nurturing innovation through multidimensional collaboration among various stakeholders (an entire book chapter explains this topic, the multi-i collaboration in depth).

Mariana Mazzucato's research on the role of the state and mission-driven innovation highlights the importance of collaboration and collective action in driving transformative change.

As socio-digital entrepreneurs, we need to understand the interdependencies and interactions within ecosystems. For example, knowledge and research institutions generate cutting-edge research and technological advancements that can be commercialised by start-ups and established companies. Entrepreneurial support

2.1 The Urgency to Plan Ahead the Transition Journey

organisations provide guidance, mentorship and access to networks, helping start-ups navigate the challenges of scaling their innovations. Funding and investment sources play a vital role in fuelling the growth of start-ups and supporting their innovative ventures. Collaboration between industry players and start-ups can lead to the adoption of new technologies, the development of new markets and the creation of value for all involved.

Understanding these interdependencies allows ecosystem actors to identify opportunities for collaboration, leverage each other's strengths and address challenges collectively. It nurtures a culture of cooperation, knowledge sharing and continuous learning within the ecosystem, leading to enhanced innovation capacity and overall ecosystem resilience.

Given the interdependencies of global challenges, collaboration and cooperation among diverse interested parties are essential, they have become more crucial than ever. No single actor or sector can tackle these issues in isolation.

Specifically, the shift towards a circular or regenerative economy, with individuals, business and organisations striving to create positive impacts rather than simply acknowledging or trying to reduce negative ones, will help design systems that work in harmony with nature. Regenerative actions go beyond sustainability as they aim to restore and enhance ecosystems and communities which aligns with the triple transition approach. This involves practices such as recycling, reusing and regenerating resources, as well as restoring biodiversity and promoting community well-being. We can address global challenges and build a more sustainable and resilient future [1] by embracing a regenerative approach.

The need for profound transformations has never been more evident. We find ourselves at a crucial juncture where fairer transformative innovation is not just a choice but a necessity.

The crossing reinforced bridge resembled in Fig. 2.3 above captures the synthesis of many years of experimentation in building stronger transition bridges within organisations and innovation ecosystems.

We can bridge the gap of the 'turning point' towards the emerging systems by establishing stronger partnerships and knowledge sharing networks.

Forming communities (interdisciplinary platforms, living labs, core groups), composed of actors who share common professional interests or goals can

Fig. 2.5 Four (or n-) helix collaboration bridge

strengthen the bonds and bridges to cross the weak path from the established system to the emerging one (Fig. 2.5).

It delves into the process between established systems and the new emerging ones. By recognising the need to reinforce the transitions gaps, we lay the groundwork for developing inclusive strategies, policies, initiatives and collaborative actions.

Building networks and coalitions allows for pooling resources, expertise and influence to create a collective impact. The aim is to build stronger bridges, creating reinforced bonds between the bottom-up and top-down initiatives. This creates a more robust transition path that helps us gauge the ups and downs, the uncertainties and risks that we experience during the process of change.

This can be done through not fixed or formalised structures, but flexible and adaptable ones that promote collaborative learning, problem-solving, exchange of insights and experiences and initiate joint projects. Continuous learning processes consider learning feedback and feedforward loops, therefore the spiralling effect.

> Creating multi-i co-creative governance tornado (as we will see in Chap. 5) support the creation of knowledge and practice-sharing communities, to cultivate and expand innovation and influence. These are groups of people within the organisation or the innovation ecosystem who come together to share knowledge, expertise and practices (mutual learning from good or bad experiences) in a specific domain or area of interest.

To facilitate this transition, it is crucial to focus on institutional changes and promote innovation ecosystems. Mere technological advancements or infrastructure development alone cannot bring about comprehensive transformation. Instead, we need to prioritise social transformation, integrating the impacts of new developments and technologies into our society.

Novel aspirational and practical approaches that are able to nurture trust, co-creation, co-responsibility and mutual care should be explored, instead of proposing closed formulas. This should entail encompassing top-down, middle-round and bottom-up avenues.

2.2 Towards a Just Triple Transition: Social, Green and Digital

Global challenges and issues such as climate change, poverty, inequality and resource depletion do not exist in isolation but are intertwined and mutually influenced. For instance, climate change exacerbates poverty and inequality, while economic inequality contributes to social and environmental injustices.

Relying solely on so-called practical, pragmatic and realistic approaches will not be enough to meet the sustainable development goals within the timeframe outlined in the 2030 agenda. It will require a bold and unwavering change in our attitudes and opinions, coupled with decisive actions and a strong commitment to speeding up global efforts for decarbonisation, quality education for all, etc. Achieving a fairer triple transition (social, green and digital) towards generating new energy, water and food models cannot be accomplished by postponing actions to address global warming, poverty eradication, ecosystem preservation or the need to strike a balance between economic gains and investments in achieving universal innovation ecosystems (Serra 2014), access to Internet as another human right, universal access to quality health systems, gender equality, etc. It's crucial that we acknowledge these challenges and work together to find viable solutions.

This section echoes the new vision for change we developed in The MoskEUteers—the 'all for one, one for all' just triple transition [4]. There, we proposed courageous ecocentric alternatives to the urgent need for more human and planet-friendly actions. The idea behind is that by collaboration and alignment of our goals, we can drive transformative impact on a larger scale. This way we can detach from a destructive (for us and for the planet) egocentrism that works mainly for our own interests, to function with a broader worldview for the common good and that is more ecocentric, human-centric and planet-friendly.

Recognising the interdependencies of global challenges is crucial for addressing them effectively. Solutions that focus on a single issue or operate within silos are often inadequate and may inadvertently exacerbate other problems. A holistic and integrated approach is needed to address the complex and interrelated nature of these challenges.

After having developed a preliminary analysis of the foresight report published by the European Commission [5–9],[3] we see that all the areas covered by the report tackle broad and complex challenges. Most of them focus on the twin transitions, digital and green (climate change, renewable energy). However, they fully embrace the social dimension, thus being completely aligned with:

- the triple transition idea, of the MoskEUteers [4],
- the necessity of generating new governance models that are more collaborative, participative and involving the quadruple (or n-) helix with the more decisive enrolment of citizenship; and
- the call for action in a more collaborative and encompassed way (better alignment and governance).

The figure below synthesises the priority areas of action for European actors in the coming years (Fig. 2.6).

To tackle innovative and advanced governance issues that enable us to overcome milestones, separate approaches and to have a more systemic and a steady view, one of the keys is to anticipate and be aware of the challenges, regulations and indices and indicators that are applied to the working and implementation scope we are in.

---

[3] This is a preliminary approach (not an exhaustive list) of regulations and legislations affecting companies, public and private entities and citizenship as presented in Annex 1. This showcases how we can start deepening on these issues, which are directly related to the Eoh-for-Good's early warning and diagnostic services.

Fig. 2.6 Priority areas of action[4]

Cultivating sustainable and inclusive innovation ecosystems requires integrating the principles of the triple transition, which encompasses environmental, socio-economic sustainability and digital transformation for the common good. While the twin digital and green transitions have gained recognition through, for instance, the EU policies and funding programmes, we believe that a more holistic approach—one that fully integrates the social dimension—is necessary for a true systemic change.

This approach acknowledges the interconnectedness between these dimensions and recognises that long-term success and competitiveness depend on their balanced consideration. By understanding the interconnectedness of global issues, actors committed to change can identify leverage points and develop strategies that create positive ripple effects across multiple domains. For example, addressing

---

[4] [5–9].

2.2 Towards a Just Triple Transition: Social, Green and Digital

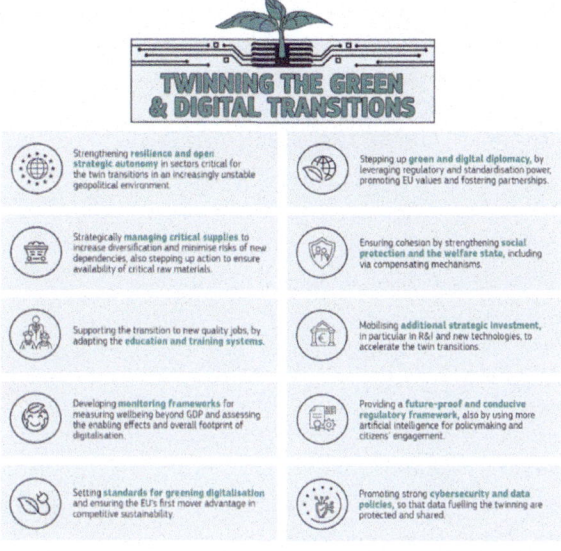

Fig. 2.7 Twinning the green and digital transitions[5]

poverty and inequality can contribute to environmental sustainability, while sustainable development practices can help alleviate poverty and enhance social well-being.

The digital transition encompasses the ongoing technological advancements that shape our societies, economies and interactions. It holds immense potential to drive efficiency, innovation and connectivity. However, without a human-centric perspective, it risks exacerbating inequality, marginalisation and privacy and security concerns.

Similarly, the green transition (Fig. 2.7):

- recognises the urgent need to address environmental challenges and build sustainable systems that minimise the negative impact on the environment;
- promotes practices that preserve natural resources, reduce greenhouse gas emissions and promote clean technologies;
- involves transitioning to renewable energy sources, adopting circular economies and promoting ecological balance.

Yet, without addressing social aspects, such as equitable access and social justice, we risk perpetuating existing disparities, and by incorporating the social dimension into the twin transition, we can promote an inclusive and equitable future. Social innovation and societal well-being need to become an integral part of our

---

[5] [10–12], accessed March 5th, 2023.

Fig. 2.8 MoskEUteers for a just triple transition[6]

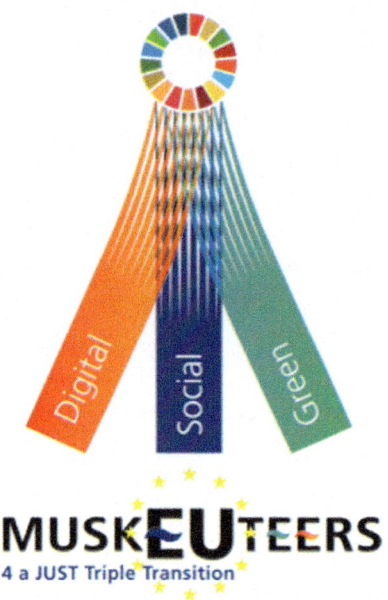

transformation efforts, ensuring that the benefits of progress are shared by all [10–12] (Fig. 2.8).

Socio-digital innovation integrates technology, social sciences and human-centric design principles to address societal challenges. Digital solutions play a vital role in promoting individuals, communities and organisations (digital platforms, data analytics and collaborative networks) to co-create innovative solutions that have a positive impact on society.[7]

By integrating sustainable practices into innovation processes, such as ecodesign, resource efficiency and circular economy principles, ecosystems can contribute to mitigating climate change and promoting a more sustainable future.

Social inclusiveness and sustainability highlight the well-being of individuals and communities within the organisation or ecosystem. This includes promoting

---

[6] [4].

[7] There are numerous examples of multi-actor co-creation. For instance, in the field of urban planning and development, multiple actors such as government entities, private business, community organisations and residents collaborate to co-create city initiatives. They work together to integrate technologies, infrastructure and services to enhance urban living and well-being, inclusion, sustainability and efficiency. This is materialised in the integration of smart sensors, data analytics and Internet of things technology to improve traffic management, enhance public safety, optimise energy consumption and create interactive citizen engagement platforms or living labs where socio-digital innovation is getting more relevant.

diversity, equal opportunities and social inclusion. Innovation ecosystems should strive to address systemic inequalities, ensure fair access to resources and opportunities and entrust underrepresented groups and marginalised communities. By fostering social cohesion and inclusivity, ecosystems can tap into a wider pool of talent, creativity and perspectives, leading to more innovative and impactful outcomes.

Forward-looking economic sustainability encompasses more than just creating economic value and ensuring long-term viability and competitiveness. It embraces a vision that integrates social and circular economy principles, regenerative approaches, resilience, quadruple helix collaboration, living labs, inclusion, access and agency.

In this transformative vision, economic sustainability goes beyond traditional notions of prosperity and instead focuses on creating shared prosperity that benefits all members of the ecosystem. It highlights the importance of rising entrepreneurship and job creation while considering the societal impact of economic activities. This involves promoting fair and inclusive business practices, supporting the development of social enterprises and ensuring equitable access to economic opportunities for marginalised communities.

References

1. Morseletto P (2020) Restorative and regenerative: exploring the concepts in the circular economy. J Ind Ecol 24(4):763–773. https://doi.org/10.1111/jiec.12987
2. Perez (2022) Portada: Carlota Perez. https://carlotaperez.org/portada/. Accessed 28 May 2023
3. Serra A, Caro-Gonzalez A, Colobrans J (2023) Collaboratories: universal innovation ecosystems in the era of transitions 17
4. Caro-Gonzalez A, Serra A, Albala X, Borges CE, Casado-Mansilla D, Colobrans J et al (2023) The Three MoskEUteers. Pushing and pursuing a one for all, all for one triple transition: social, green and digital. In: Facilitation in complexity: from creation to co-creation, from dreaming to co-dreaming, from evolution to co-evolution. https://www.researchgate.net/publication/363663350_The_Three_MoskEUteers_Pushing_and_pursuing_a_One_for_all_All_for_one_triple_transition_social_green_and_digital. Accessed 8 June 2023
5. European Commission (2023a) Communication from the commission to the European parliament and the council: sustainability and people's wellbeing at the heart of Europe's open strategic autonomy (No. COM (2023) 376 final). Brussels, p 21. https://doi.org/10.1163/2210-7975_HRD-4679-0058
6. European Commission (2023b) European pillar of social rights—building a fairer and more inclusive European Union. https://ec.europa.eu/social/main.jsp?catId=1226&langId=en. Accessed 19 June 2023
7. European Commission (2023c) Green education initiatives| European education area. https://education.ec.europa.eu/node/1745. Accessed 14 July 2023
8. European Commission (2023d) Social innovation and transnational cooperation| European social fund plus. https://ec.europa.eu/european-social-fund-plus/en/social-innovation-and-transnational-cooperation. Accessed 21 July 2023
9. European Commission (2023e) Sustainability-related disclosure in the financial services sector. https://finance.ec.europa.eu/sustainable-finance/disclosures/sustainability-related-disclosure-financial-services-sector_en. Accessed 1 July 2023

10. European Commission (2022a).2022 strategic foresight report. Twinning the green and digital transitions in the new geopolitical context. Brussels, p 17. https://knowledge4policy.ec.europa.eu/publication/2022-strategic-foresight-report-%E2%80%9Ctwinning-green-digital-transitions-new-geopolitical_en
11. European Commission (2022b) Recomendación del consejo relativa al aprendizaje para la sostenibilidad medioambiental (No. COM (2022) 11 final). Brussels. https://eur-lex.europa.eu/legal-content/ES/TXT/PDF/?uri=CELEX:52022DC0011&from=EN
12. European Commission (2022c, December 14) Intergovernmental panel on climate change (IPCC). https://research-and-innovation.ec.europa.eu/research-area/environment/climate-change-science/intergovernmental-panel-climate-change-ipcc_en. Accessed 2 July 2023

Open Access This chapter is licensed under the terms of the Creative Commons Attribution 4.0 International License (http://creativecommons.org/licenses/by/4.0/), which permits use, sharing, adaptation, distribution and reproduction in any medium or format, as long as you give appropriate credit to the original author(s) and the source, provide a link to the Creative Commons license and indicate if changes were made.

The images or other third party material in this chapter are included in the chapter's Creative Commons license, unless indicated otherwise in a credit line to the material. If material is not included in the chapter's Creative Commons license and your intended use is not permitted by statutory regulation or exceeds the permitted use, you will need to obtain permission directly from the copyright holder.

Chapter 3
Tailoring Transformative Governance for the Common Good

This chapter delves on the exploration of key approaches and principles that underpin transformative governance. These act as guiding values, ensuring that decision-making processes and policy formulation remain rooted in ethics, social justice and sustainability. By adhering to these principles, transformative governance has the potential to truly create a positive impact on society both from a social and societal approach to impact.

It explores the interconnectedness of social, green and digital considerations rising awareness on the need to address these dimensions in a balanced and integrated manner. The travel arrow pointing towards the common good serves as a directional guide for transformative governance creating a more equitable and sustainable society.

The aspiration is to help us navigate the challenges of transition with purpose and responsibility, ensuring positive outcomes for society through the alignment of our actions with the common good.

3.1 A Compass in a Nutshell—The Navigating Tool

This section introduces and illustrates the different elements of a compass. The use of metaphors helps us explain complex processes of change and participation with easier to understand images.

I have reflected on the match between the elements of a conventional compass, as a navigating tool, with the features of our processes of change in a company, a public entity, a social enterprise. This is reflected in the figure and the table below for organisations and/or in innovation systems to consider when navigating change and while designing solutions to solve small or bigger problems detected.

The compass is introduced as a guiding tool for transformative governance, providing a sense of direction and purpose as we navigate the complexities of transition. It serves as a valuable resource for decision-makers and stakeholders, helping them align their efforts with the common good.

Configured as a unique 360° navigating tool, it helps us drive socio-digital innovation and collaborative and transformative quadruple (or n-) helix co-creation dynamics which:

1. integrates system thinking from a human-centric and user-centric design;
2. fosters organisational and ecosystem development in pursuit of the common good;
3. recognises the multidimensional nature of challenges and the interconnectedness of different constituents, driving institutional or ecosystem change as a collective endeavour. This involves the integration of several elements or dimensions that start with an 'i' engaging actors at all levels; and
4. equips organisations and innovation ecosystems to design:

   (a) short-medium and long-term roadmaps and frameworks of intervention and theories of change;
   (b) identify intra-entrepreneurs and provide ad hoc accompanying services and capacity building to facilitate 'sherpas', agents of change, innovators for companies and innovation ecosystems to reach their heights; and
   (c) define positive impacts monitoring mechanisms to address positive/regenerative societal impacts.

> We need more cohesive and interconnected manner for organisations and ecosystems to navigate complex paths to change. Like in a compass, the challenges of the triple transition and the urgency to cultivate new ways of collaboration, new governance (systemic, socio-digital, etc.) entail several elements to work together. The lessons learned and shared in this book can help deploy in-house and in-context governance mechanisms and processes pointing towards the common good.

The Eoh-for-Good holistic approach fosters synergy, knowledge exchange and collective action, driving sustainable and impactful processes of change within institutions and innovation ecosystems. By adopting this comprehensive approach, actors committed to change can harness the full potential of innovation driving into fairer, sustainable and inclusive solutions and outcomes (Fig. 3.1).

The features of a modern compass are presented in the following table with the corresponding Eoh-for-Good attributes in the second column (Table 3.1).

3.2 Grounding Approaches and Principles for a Just Triple Transition

Effective management involves more than simply providing overly aspirational or generic responses, it entails posing critical questions and promptly implementing tangible initiatives.

3.2 Grounding Approaches and Principles ...   25

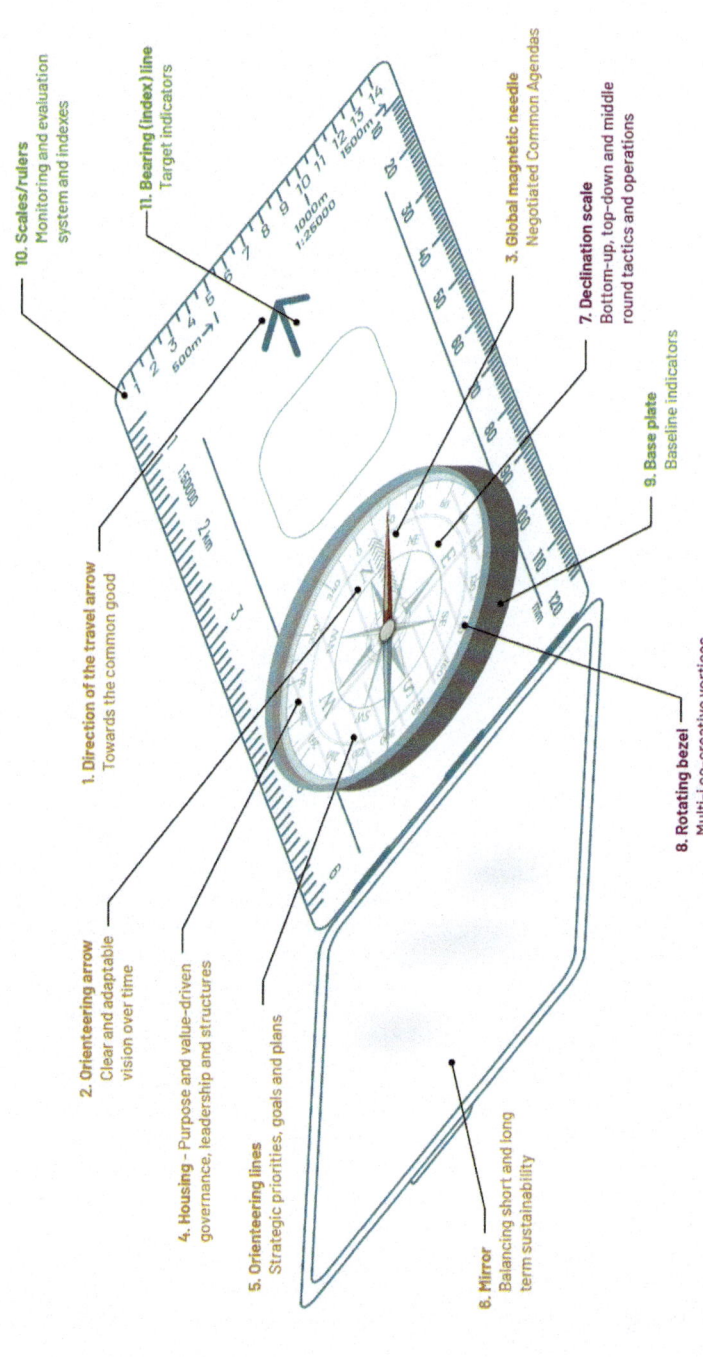

Fig. 3.1 Eoh-for-good compass for social transformation

Table 3.1 Eoh-for-good compass features

Features of a modern compass[1]	Features of the eoh-for-good compass
Chap. 4: *Grounding approaches and principles for a just triple transition*	
1. *Direction of travel arrow* Marked on the base plate. Tells you which direction to point the compass when you're taking or following a bearing	*Towards the common good* In the eoh-for-good compass, this is aligned with some dimensions of the inner development goals framework *Inner compass*: Having a deeply felt sense of responsibility and commitment to values and purposes relating to the good of the whole (Inner Development Goals n.d.) *Integrity and authenticity*: A commitment and ability to act with sincerity, honesty and integrity
2. *Orienteering arrow* Marked on the floor of the housing. It rotates with the housing when the dial is turned. You use it to orient a compass to a map. It has an outline shaped to exactly fit the magnetised end of the needle	*Clear and adaptable vision over time*: A well-defined and inspiring vision that outlines the desired future It serves as a guiding principle for decision-making and action
3. *Global magnetic needle*[2] Magnetised piece of metal that has one end painted red to indicate North/South*. It sits on a fine point that is nearly frictionless, so it rotates freely when the compass is held fairly level and steady	*Negotiated common good agendas* Alignment of interests and needs transiting from an egocentric towards an ecocentric approach Shared transformative agendas, outcomes and impacts need to be co-created, developed, implemented and continuously renegotiated and redefined with internal and external entrepreneurs, quality agencies, policy makers, social and business-driven innovators, companies (SMEs and corporations) and cutting-edge technology centres
4. *Housing* The main part of the compass. It is a round plastic container filled with liquid and has the compass needle inside. A bubble of air in the housing liquid is useful for making sure you are holding the compass fairly level	*Purpose and value-driven governance, leadership, structures and infrastructures* encompass the digital platforms, technological systems, and networks that lay the foundation for innovation and collaboration. They facilitate the smooth exchange of information, resources, and ideas among diverse actors, enabling co-creation and knowledge sharing. Leadership and governance, characterised by conscious, human-centred, and compassionate approaches, provide direction, guidance and support for change processes and strategies. They ensure accountability, effective decision-making and coordinated efforts in driving positive outcomes

(continued)

---

[1] *Source* Anatomy of a hiking compass | Hike Navigation.

[2] Standard compasses only work in one hemisphere, so they are either northern or southern hemisphere specific. More expensive models can work in both hemispheres and therefore function well worldwide. This is worth keeping in mind if you are travelling to other countries.

3.2 Grounding Approaches and Principles …

Table 3.1 (continued)

Features of a modern compass[1]	Features of the eoh-for-good compass
5. *Orienteering lines* Series of parallel lines marked on the floor of the housing and on the base plate. Correctly aligning these with the north–south lines on a map aligns your orienting arrow with north	*Strategic priorities*: Key focus areas and priorities that align with the vision and goals, ensuring that efforts are directed towards the most critical areas of change *Goals and objectives*: Specific and measurable goals and objectives that outline the desired outcomes and milestones to be achieved along the journey of change *Action plans*: Detailed plans and initiatives that outline the specific actions, tasks and timelines needed to implement the strategy and drive the desired changes
6. *Mirror* Let's you see the compass face and distant objects at the same time (using the sight) which provides more accurate readings. The mirror can also double as an emergency signalling aid *Sight*: Can be helpful to aim more precisely at a distant landmark, especially in open terrain *Magnifier*: For more detailed reading of map features	*Balancing short-term gains with long-term sustainability* in the processes of change in institutions and innovation ecosystems involves managing immediate benefits while ensuring actions align with long-term sustainable goals. It requires making strategic choices that prioritise both short-term success and the preservation of resources, environmental responsibility and social impact for lasting positive transformations
Chap. 6: *Embarking on transformative transition processes*	
7. *Declination scale* Used to orient the compass in an area with known declination and is also used for easily adding or subtracting the known declination in your area of travel	*Eoh-for-good joystick* triggers bottom-up, top-down and middle-round collaborative initiatives and processes within or across units, departments, divisions, institutions or ecosystems How does it work?
8. *Rotating bezel* Also called the 'azimuth ring', this outer circle has 360-degree markings. You hold the dial and rotate it to rotate the entire housing	*Multi-i co-creative vortexes* refer to specifically processes designed to facilitate collaboration, creativity and co-creation among diverse stakeholders. These processes materialise in spaces, such as living labs, innovation hubs, co-working spaces and digital platforms, that serve as platforms where individuals from academia, industry, government and civil society can come together
Chap. 7: *Designing ad hoc monitoring system*	
9. *Base plate* Hard, transparent, flat surface on which the rest of the compass is mounted. It has a ruler on its edges for measuring distances on maps. Its edge is straight and is used for taking bearings and transferring them to your map	*Baseline indicators* refer to the initial set of measurable variables or parameters used to establish a reference point or starting point for monitoring and evaluating progress. These indicators serve as a benchmark to measure the current state or condition of an organisation, system, project or initiative before any interventions or changes occur. Baseline indicators provide a basis for comparison, enabling responsible actors to assess the effectiveness and impact of interventions by comparing subsequent measurements against the initial baseline data

(continued)

Table 3.1 (continued)

Features of a modern compass[1]	Features of the eoh-for-good compass
10. *Scales/Rulers* Each edge of a compass may have different rulers for use with different map scales. In Australia, a compass with scales of 1:25 k and 1:50 k will work best. Use the scale that corresponds with your map's scale to determine distances	*Monitoring and evaluation system for deployment of outcomes, results and impacts*: Regular monitoring and evaluation of progress and outcomes to track the effectiveness of the strategy, identify areas for improvement, and make necessary adjustments along the way Results, outcomes, and impacts in processes of change within organisations and innovation ecosystems refer to the tangible and intangible changes that occur because of transformative efforts. Results indicate specific outputs, outcomes signify broader changes or achievements, and impacts denote the long-term effects and benefits generated, including social, economic and environmental dimensions
11. *Bearing (index) line* Located directly above the bezel, it's also called a 'read bearing here' mark	*Target indicators*: Key performance indicators (KPIs) to plan and to track the effectiveness of the strategy, identify areas for improvement and make necessary adjustments along the way

3.2.1 The Travel Arrow Pointing Towards the Common Good

Eoh-for-Good is grounded on the common good as a framework that serves for understanding the well-being of all, including specially the most vulnerable, marginalised or at risk of exclusion.

There are three elements of the compass, (1) the global magnetic needle, (2) the direction of travel arrow and (2) the orienteering arrow, that point into the right direction to help us navigate when we are in open terrain. They help us to find our path and keep on track on our way forward (Table 3.2).

Note that three underlying dimensions that start with an 'i': **inclusion, innovation and impact** drive our 360° vision, approaches and principles to change. These are the first three 'i's of the model that permeate everything action line.

The common good is valued above all other considerations, even when there is disagreement about how best to achieve it. It embodies the idea that, beyond individual interests, there are shared values and objectives that benefit society. Furthermore, it involves promoting social justice by addressing inequalities in a systemic manner, considering the broader societal positive impacts.

But what do we understand by common good? The concept of the common good is an idea that aims to promote a common understanding of what constitutes good for all. However, although the idea of the common good has been well-established in moral philosophy, there is no consensus on its meaning, especially when a single formula cannot be applied to very diverse contexts.

3.2 Grounding Approaches and Principles … 29

Table 3.2 Elements 1, 2 and 3 of the compass

Element of a compass		Eoh-for-good features
	1. *Direction of travel arrow* Marked on the base plate. Tells you which direction to point the compass when you're taking or following a bearing	*Towards the common good* In the eoh-for-good Compass this is aligned with some dimensions of the Inner development goals framework *Inner compass*: Having a deeply felt sense of responsibility and commitment to values and purposes relating to the good of the whole *Integrity and authenticity*: A commitment and ability to act with sincerity, honesty and integrity
	2. *Orienteering arrow* Marked on the floor of the housing. It rotates with the housing when the dial is turned. You use it to orient a compass to a map. It has an outline shaped to exactly fit the magnetised end of the needle	*Clear and adaptable vision over time*: A well-defined and inspiring vision that outlines the desired future It serves as a guiding principle for decision-making and action
	3. *Global magnetic needle* Magnetised piece of metal that has one end painted red to indicate North/South*. It sits on a fine point that is nearly frictionless, so it rotates freely when the compass is held fairly level and steady	*Negotiated common good agendas* Alignment of interests and needs transiting from an egocentric towards an ecocentric approach Shared transformative agendas, outcomes and impacts need to be co-created, developed, implemented and continuously renegotiated and redefined with internal and external entrepreneurs, quality agencies, policy makers, social and business-driven innovators, companies (SMEs and corporations) and cutting-edge technology centres

To make sense out of this concept and apply it in practical terms, we need to understand how it relates to other interconnected concepts and frameworks, such as justice (which requires achieving equal opportunities and outcomes for all), efficiency (which aims to achieve the best possible long-term outcomes with limited resources) and inclusion (what implies leaving no one behind, what usually implies higher levels of public investment).

The common good is understood as an ideal moral measure [1, 2] or as a model that points to a state of affairs, a world in which all the conditions are in place (such as access to good education and health care, safe housing, fair wages, the ability and access to participate in political and cultural life, etc.) to allow everyone to reach their full potential as persons and communities. This provides us with a compass to:

- define priorities[3] and shared agendas[4] where we collectively design at different levels (local or global) a common ground from where everybody's voice, agency and participation in decision-making can be taken into consideration (e.g. a theory of change defined with the participation of all actors);
- delimit a moral measure—a tool that we can use to evaluate whether our choices, policies and institutions align with the jointly defined ideal path; and
- avoid anyone claiming to know precisely what is suitable for all persons without having into account the many different facets of reality. Individual or small group attempts to specify the common good can fall short of being truly common or genuinely.

Why do it collectively? Humans tend to think that we can judge and decide everything on our own. However, our moral vision can be mistaken, biased or limited by our own experiences, prejudices, backgrounds and more. To improve, we should question and challenge our own beliefs and values. We need to be humble, open-minded and receptive to new ideas and challenges. This is important not only because this way we may be able to improve ourselves but also better understand the world around us.

The common good can only be established if we are able to imagine and interpret the wide range of moral commitments, principles and obligations that influence us in a collective manner. This is clearly perceived with the misuse of resources and deterioration of the planet, for instance.

The common good is not a single objective state, but a way to unite our best intentions to create something that benefits all of us. The idea of the common good rests on a pretty robust optimism about human beings. It does not deny that social conflict will happen, but we assume that since humanity is built for community within a wider ecosystem and context, we are willing to treat each other with goodwill and collaborate in creating the common good also for the planet. This assumed optimism about human beings can be hard to sustain when there is so much evidence of ill-well and such widespread forms of violence, inequality or harm to the planet.

However, Eoh-for-Good aims precisely to inject a dose of positivism, not meaning naive optimism, but conscious concern to improve things harbouring these principles, both ethically and for the sake of effectiveness in its impact. It is an attempt to create a community of people or groups who are working towards a shared goal and a common good. It exists whenever there are projects or initiatives in existence, where all members of society can participate equally, regardless of their location, background, status and culture.

---

[3] https://www.camarabilbao.com/corporativo/bizkaia-2050-vision-futuro-territorio-202204281232/, accessed November 28, 2022.

[4] https://sdgs.un.org/goals, accessed October 28, 2022.

3.2 Grounding Approaches and Principles …

This just transformative innovation should base its work dynamics in the following interdependent approaches and principles underlying the common good, which are reflected in the global agenda[5] and European priorities, policies and principles[6] and many other local/regional and national policies, strategic plans, initiatives and roadmaps.

3.2.2 Approaches

The following are our main four approaches to change:

1. **Purpose and value-driven organisational governance, structures and infrastructures**

To expand the scale and effectiveness of organisations and innovation ecosystems, it is crucial to build innovative and organisational governance, structures and infrastructures and a solid housing from where to address the transitions (Table 3.3).

In the Eoh-for-Good approach, institutional change requires leadership and governance that prioritise transparency and accountability for the common good. It involves fostering a culture of integrity, where ethical considerations are embedded in decision-making processes and the allocation of resources. Ethical institutional change promotes inclusive and participatory governance, ensuring that diverse voices are heard and represented. Ethical responsibility entails conducting business in a manner that:

- respects human rights;
- promotes fair trade practices;
- embraces entrepreneurship and inclusive business models that address social and environmental challenges while generating economic value; and
- considers the social-economic-political and environmental (also unexpected or undesired) impacts of products and services.

---

[5] From the United Nations, we note that the common good is driven through the Platform for Partnerships as a global registry of voluntary commitments and multi-stakeholder partnerships in support of the implementation of the sustainable development goals (SDGs), and through various United Nations conferences and thematic action networks, such as the United Nations Conference on Oceans, the Conference of Small Island Developing States, the United Nations Conference on Sustainable Transport, the Rio+20 Conference and others. To achieve this common good, all interested parties are encouraged to register their partnerships and voluntary commitments that support the implementation of the SDGs.

[6] Another source we can turn to when looking at Europe is the 20 principles of the European Pillar of Social Rights, which aim to consolidate a strong, fair, inclusive and opportunity-filled social Europe. The Commission has already presented several actions based on each of the Pillar's principles, and new ones are planned to further strengthen social rights in the EU. These principles are implemented on the basis of the European Pillar of Social Rights Action Plan.

Table 3.3 Element 4 of the compass

Element of a compass		Eoh-for-good features
	4. *Housing* The main part of the compass. It is a round plastic container filled with liquid and has the compass needle inside. A bubble of air in the housing liquid is useful for making sure you are holding the compass fairly level	*Purpose and value-driven approaches and principles to governance, leadership, structures and infrastructures* encompass the digital platforms, technological systems, and networks that lay the foundation for innovation and collaboration. They facilitate the smooth exchange of information, resources, and ideas among diverse actors, enabling co-creation and knowledge sharing. Leadership and governance, characterised by conscious, human-centred, and compassionate approaches, provide direction, guidance, and support for change processes and strategies. They ensure accountability, effective decision-making, and coordinated efforts in driving positive outcomes

In many cases, institutions have become rigid, bureaucratic or resistant to change, hindering their ability to address emerging challenges and seize new opportunities. Recognising these limitations is the first step towards driving institutional change.

By critically examining existing institutional practices, leaders and key players can identify areas that require transformation.

This involves evaluating the effectiveness of decision-making processes, the responsiveness to changing socio-political, cultural and market dynamics with the consequent alignment with societal needs and expectations. Understanding the shortcomings of current institutions provides the impetus for change and creates a sense of urgency to drive the necessary transformations.

Ethical responsibility requires actors committed to change to reflect on the consequences of their decisions and actions. It demands a shift from a purely profit-driven mindset to one that considers the broader implications and externalities of actions (e.g. the potential impacts on marginalised communities, future generations and the planet as a whole).

Eoh-for-Good is driven by the recognition that existing systems and structures may no longer be effective or aligned with the needs of a rapidly evolving world.

3.2 Grounding Approaches and Principles ...

2. Foster conscious, human-centred, compassionate leadership

 Conscious, human-centred and compassionate leadership[7] are based on self-awareness, empathy and a deep understanding of the needs and aspirations of individuals and communities. Leaders who embrace these principles prioritise the well-being and growth of their team members, promoting a supportive and inclusive work environment. They lead with integrity, authenticity and a genuine concern for the greater good. By embracing these principles, leaders can create a supportive and inclusive work environment where all parties are understood and valued.

Several renowned management theorists, leadership experts and practitioners have greatly contributed to develop these principles and applied them to organisational development. I can recall the work of four, although it goes without saying that there are many others:

- Daniel Goleman in his book 'Primal Leadership: Unleashing the Power of Emotional Intelligence' explored how leaders can cultivate emotional intelligence, empathy and self-awareness to create positive organisational climates and drive sustainable success. This psychologist highlights the importance of compassionate leadership in raising employee well-being and engagement [3].
- Ackoff's systems approach [4–6] underlines the importance of aligning an organisation's purpose with its strategy and operations. Entrepreneurs can cultivate a sense of ownership and accountability by developing a clear and compelling vision that inspires and motivates employees. For example, a leader that communicates the organisation's mission and values in an effective and transparent way can help everyone know how their individual efforts contribute to the collective goals.
- Schein [7] focuses on creating a culture of continuous learning and growth. By fostering a supportive work environment, leaders encourage individuals to contribute their best and strive for collective excellence. Leaders can establish mentorship programmes, where experienced employees guide and support others in their professional development. This not only enhances individual growth and strengthens the overall performance of the organisation.

---

[7] Authors who have developed and applied these principles to organisational development are:

Simon Sinek, a renowned author, speaker and leadership expert known for his work on inspiring leaders to create a sense of purpose and engage their teams. In his book 'Leaders Eat Last: Why Some Teams Pull Together and Others Don't.' Fred Kofman, a leadership and organisational development expert, through his book 'Conscious Business: How to Build Value Through Values' explores the principles of conscious leadership, focusing on personal growth, emotional intelligence and ethical decision-making. Kofman advocates for creating institutions that nurture genuine connections, purpose-driven work and compassionate leadership.

- Drucker [8] people-centred principles highlight the importance of purpose-driven leadership and organisational effectiveness. To align their strategies with the organisation's objectives, leaders need to set clear goals and measuring progress through the implementation of feedback mechanisms and regular performance evaluations to identify areas for growth and development. Besides, leaders create an environment that nurtures creativity and continuous improvement by prioritising knowledge work and encouraging innovation.

Our approach is to promote conscious, human-centred, compassionate leadership, which is crucial to understand and empathise with all parties involved. This means avoiding judgement, one-sided approach, biased, blaming or punishing. Leaders embracing love and compassion towards all parties listen to all perspectives, considering the entire system and having compassion for all its parts. This leads to make informed decisions that consider the well-being and success of every party involved and to benefit the common good.

In practice, this might involve conducting inclusive meetings where different voices are heard and valued. Leaders promote open and honest communication, creating a safe space for sharing diverse opinions and ideas. By seeking to understand the needs and concerns of all parties, leaders can make informed decisions that consider the well-being and success of every party involved.

3. **Systemic Change**

In the pursuit of addressing global challenges and achieving the just triple transition, it is essential to balance individual interests with the collective well-being. While individuals and organisations have their own goals and objectives, it is crucial to consider the broader and more systemic societal and environmental implications of our actions. This requires considering the wider picture.

Looking at reality with a more holistic and multidimensional approach considers complexity, collaboration and alignment with the common good. System thinking is a powerful tool for understanding the interconnectedness and interdependencies within complex systems [6]. It motivates actors committed to change to take a holistic view and consider the broader context in which innovation takes place. From a systemic approach, actors can develop innovative solutions that contribute to sustainable, inclusive and just change with an audacious aspiration to transition towards a future state that offers fundamentally improved outcomes for individuals, communities and the environment.

While systems innovation and systemic change are related concepts, they differ in scope and focus. Systems innovation focuses on improving specific elements within a system, while systemic change aims for broader, transformative shifts that encompass the entire system and its interconnections.

Eoh-for-Good approach focuses on this second avenue of systemic change as this aims to address the root causes of problems to then design lasting, sustainable transformations across multiple interconnected systems.

It entails broader and more transformative shifts that go beyond incremental improvements within a specific system, rethinking and restructuring the entire system itself, including its underlying structures, processes and relationships. This often involves shifts in norms, values, mindsets, policies and implementations to bring about comprehensive change.

This holistic approach encourages synergy, knowledge exchange and collective action, driving sustainable and impactful innovation within the ecosystem.

> Actors committed to change can unlock the potential of intra-entrepreneurship, drive institutional transformation and create a more sustainable, competitive and inclusive future for all by embracing a holistic and integrated approach.

Public institutions play a crucial role in shaping the social, economic and political landscape. They provide an important framework within which individuals and organisations operate, influencing behaviours, norms and outcomes.

4. **Human-centred design (HCD)**

It is a people-focused approach used in design and management to tackle problems by involving individuals' perspectives throughout the entire process of innovation. It starts with observing the problem in its context, brainstorming ideas from different viewpoints, conceptualising solutions, creating prototypes, and iteratively testing them before implementing the final solution. HCD goes beyond simply documenting perspectives; it actively engages collaborators, including their diverse needs, desires and viewpoints, during participatory processes. The initial stages involve immersing oneself in the problem and community,[8] observing and framing the context. By designing for real people and their everyday realities, HCD helps identify and efficiently solve the right problems using local resources and minimal means, ensuring the seamless transition from ideas to concrete solutions.

---

[8] For this is very interesting the techno-anthropology approach, which is an interdisciplinary field that explores the dynamic relationship between technology and human societies, combining anthropological methods with technological research to examine how innovations impact cultures and human behaviour. In practical terms, techno-anthropology involves conducting empirical research to understand how technology and society intersect. This may include ethnographic studies, surveys, interviews, and observations to analyse how people interact with technology, how it shapes their behaviours', and how cultural norms influence the adoption and use of technology. The insights gained from such research help inform the design, development and implementation of technologies that align better with human needs and values. Techno-anthropologists also assess the social and ethical implications of technological advancements to promote responsible innovation and address potential risks and challenges that arise from the integration of technology into everyday life.

5. Continuous Learning and Adaptation

As mentioned before, a commitment to ongoing learning, flexibility and adaptability, will help us recognise that change is a ´dynamic process that requires adjustments based on new information, emerging challenges and evolving contexts. Experimentation and learning become key components of the transition process. We must be open to exploring new approaches, testing innovative solutions and learning from failures. By embracing a culture of experimentation, we can iterate and refine our strategies, adapting to rapidly changing circumstances and emerging insights.

Embracing slow and inflexible management hampers innovation, development and the search for solutions to intricate problems.

3.2.3 Principles

The **leave no one behind** principle means **ending extreme poverty in all its forms and reducing inequalities among both individuals (vertical) and groups (horizontal)**. It entails a holistic view of the situation in each context that includes all persons from different backgrounds and sectors, prioritising and fast-tracking actions for the poorest and most marginalised people. This is known as progressive universalism [9], which aim is to create synergies that tend to avoid the constant marginalisation of vulnerable groups.

With the leaving no one behind principle, the European Commission's aims to provide cohesive funding to all regions equally and to ensure that no region is left behind, as this would create a negative and stable synergy over time in the unequal development of the European regions.

Inclusion and diversity require creating an environment that values and celebrates differences by actively seeking input from individuals with different backgrounds, experiences and expertise. Special attention should be paid to actively include individuals from underrepresented groups, such as women, minorities and individuals with disabilities.

Inclusivity should extend beyond individual representation to encompass equitable access to resources, opportunities and networks. Removing systemic barriers and biases is crucial for ensuring that all individuals, regardless of their background or identity, have equal chances to participate, contribute and succeed within the ecosystem.

Embracing diversity and inclusivity in all its forms ensures that the voices and needs of all interested parties[9] are considered. This entails:

---

[9] I will avoid using the term stakeholders for its negative connotations with native communities.

- to establish appropriate inclusive policies and practices, targeted programmes and mentorship initiatives for these individuals to participate with real voice and agency,[10] as promoted by ref. [10],
- to engage with local communities to understanding their needs and aspirations which is crucial for creating sustainable and inclusive innovation ecosystems. Collaboration between ecosystem actors and local communities can lead to the development of solutions that address pressing social and environmental challenges, ensuring that innovation benefits all members of society;
- to entrust and restitute underrepresented groups and marginalised communities, leading to more informed and equitable outcomes;
- to address systemic inequalities and barriers to entry that hinder equal participation and access involve identifying and dismantling structural and cultural barriers that disproportionately affect certain groups;
- to raise awareness and promote cultural change, essential to address biases and stereotypes that limit opportunities for marginalised individuals; and
- to leverage the playing field and providing opportunities for success.

This is linked with the 'i's related to personal traits (interpersonal, intercultural, intersectional and intergenerational) to foster creativity, enhancing problem-solving capabilities, knowledge creation and more innovative decision-making processes.

> By bringing together individuals from diverse backgrounds, cultures, ages and disciplines, individuals, organisations, communities and innovation ecosystems can tap into a wide range of ideas, insights, perspectives, experiences, knowledge and solutions. This way they are better equipped to enhance collaboration, attract and retain diverse talent and cultivate a more vibrant and resilient just sustainable change.

From egocentric to ecocentred[11] principle. To embark on this just triple transition, we must shift from egocentric to ecocentric behaviour. Our current systems often prioritise self-interest, competition and short-term gains. However, an ecocentric approach places the well-being of the planet and its inhabitants at the forefront, cultivating collaboration, collective

---

[10] Agency within the sustainable development approach refers to the capacity of individuals and communities to act, make choices and shape their own lives in pursuit of well-being. It encompasses personal freedoms, empowerment and collective action for social change. It emphasises the removal of barriers and the provision of opportunities to enable people to exercise their agency and contribute to sustainable development, while also recognising the responsibility to consider the welfare of others and the environment.

[11] Ecocentrism is an ethical and philosophical perspective that places intrinsic value and importance on the natural world and ecosystems. It contrasts with anthropocentrism, which prioritises human interests and well-being above all else. Ecocentrism considers the Earth and its ecosystems as interconnected and interdependent systems, where every component plays a vital role. It advocates for ecological integrity, conservation and sustainability. From an ecocentric perspective, the well-being of the environment and all living beings, including humans, is intertwined and the health of ecosystems is essential for long-term survival and thriving.

responsibility and long-term sustainability. Individual actions (regardless of the actor they come from) will potentially have more excellent value and impact if they enter a collaborative cycle.

> Ecocentric behaviour acknowledges the interconnectedness of all living beings and ecosystems. It recognises that our actions have far-reaching consequences and that a thriving future requires a balance between human needs and planetary limits. By embracing ecocentric principles, we can drive transformative change that is regenerative, inclusive and socially just.

Any initiative or programming that is launched must help at least two or more people to avoid falling into selfish or self-centred approaches. This means a shift in mentality towards more inclusive and less partial or serving only self-interest approaches. All living beings teach us to operate harmoniously in diversity offering us the opportunity to learn, grow and adapt. This perspective forces us to experiment with more co-creative options and the impact is felt by more fields and individuals.

No harm principle. Linked with the previous principle, the proposed progress should never harm any living being, not only for human beings but for all creatures in the ecosystem, taking special care of the environmental balance and beyond, in line with the regenerative movement: if we have the power to damage or destroying, we have also the possibility of restoring. This principle is particularly important to propose solutions to make this planet a better place to live in line with the emerging theoretical currents above mentioned (e.g. ecocentrism), which alerts on the damage we are doing to the biosphere in pursuit of human benefit.

The European Commission (EC) has established the principle of no harm to the Recovery and Resilience Plan, thus shielding the six environmental objectives as dictated by art.17 of the Taxonomy Regulation.

Win–win principle. The aim is to amplify the impact of individual endeavours, surpassing their isolated contributions, pooling together their resources, expertise and influence to achieve collective impact.

We are only stronger together! We need to co-dream, co-create and co-evolve together! (ex. The amazing experiment of massive co-creation is the EUvsVirus

> Ecocentrism promotes a holistic approach to environmental ethics, recognising the intrinsic value and inherent rights of non-human entities, such as plants, animals, rivers and mountains. It underlines the need for stewardship, responsible resource management and ecological restoration to maintain the balance and resilience of ecosystems.
> This perspective calls for a shift in human attitudes and behaviours towards more sustainable and harmonious interactions with the natural world. It encourages reevaluating our relationship with nature, recognising the Earth's limits and adopting practices that promote ecological well-being and long-term sustainability.

phenomenon[12]) [11]. This relationship is effective when all parties involved in contributing to a project or initiative find the benefits that each actor will gain from the situation, the product, service or intervention.

Shared transformative agendas, outcomes and impacts need to be co-created, developed, implemented and continuously renegotiated and redefined with internal and external entrepreneurs, quality agencies, policy makers, social and business-driven innovators, companies (SMEs and corporations) and cutting-edge technology centres.

For this to happen, it is necessary to create collaborative synergies to motivate co-creation as a more recurrent process when planning projects. This is embodied in the quadruple helix dynamics of actors from the different axe involved, where their diverse contributions are key.

> A shared vision should be communicated in a compelling and inclusive manner, ensuring that all actors understand and connect with the purpose of the change initiative. It should provide a clear picture of the desired outcomes and the positive impact that the change will have on the institution or the innovation ecosystem, its key players and the broader society. A shared vision creates a sense of purpose and fosters a collective commitment to driving institutional transformation.

We have provided a comprehensive exploration of the grounds that nurture systemic entrepreneurial change. By synthesising insights from influential authors and incorporating key concepts from various disciplines, we have uncovered a wealth of principles that can guide entrepreneurs in driving positive transformation. These principles offer a solid foundation for navigating the complexities of entrepreneurship, encouraging innovation and creating meaningful, conscious, compassionate and purpose-driven impact.

By connecting these principles to the just triple transition and the attributes of the common good, entrepreneurs have the opportunity to make a significant contribution. Through personal development and inner transformation, entrepreneurs can cultivate the self-awareness, resilience and purpose necessary to navigate the challenges of societal transitions. Professional development and organisational excellence that enable entrepreneurs to build high-performing teams and organisations can contribute greatly to societal well-being.

Institutional development and leadership entrust entrepreneurs to align their organisations' purpose with societal needs and promote a culture of trust and collaboration. By doing so, entrepreneurs can drive change that is responsive to the broader goals of the just triple transition and the common good. Lastly, innovation ecosystem development and collaborative leadership allow entrepreneurs to leverage the collective intelligence and resources of diverse interested parties, leading to transformative change on a systemic level.

---

[12] https://www.euvsvirus.org/.

As you embark on your own entrepreneurial journey, let these principles guide you towards becoming an agent of change. As you continue reading, you will uncover the 'how's to it', the navigating tool and approaches to drive positive change in your entrepreneurial endeavours. So, let's move forward building upon these principles and each one's learnings to develop a meaningful and lasting impact.

3.3 Long-Term Goal-Aligned Alternatives

Always keep your vision alive and stay focused, regardless of the challenges that come your way, and you'll undoubtedly achieve success!

While short-term gains and immediate results are important for the success of organisations and ecosystems, it is crucial to balance them with long-term sustainability. This requires several shifts from (Table 3.4):

(a) an egocentric to an ecocentric behaviour, as explained before;
(b) a mainly profit-driven approach to a more holistic and future-oriented perspective; and
(c) the ability to develop and uphold dedication towards long-term visions that are interconnected with the broader context.

This way socio-digital innovations should prioritise the creation of shared value, considering the interests of all actors involved.

Balancing short-term gains with long-term socio-economic and environmental inclusive sustainability, transformative dynamics necessitates:

1. **Identify and address potential trade-offs between short-term profitability and long-term sustainability goals**. This may involve making strategic investments, creating supportive policies and regulations and promoting a culture of responsible innovation;

Table 3.4 Element 5 of the compass

Element of a compass		Eoh-for-good features
	5. *Orienteering lines* Series of parallel lines marked on the floor of the housing and on the base plate. Correctly aligning these with the north–south lines on a map aligns your orienting arrow with north	*Strategic priorities*: Key focus areas and priorities that align with the vision and goals, ensuring that efforts are directed towards the most critical areas of change *Goals and objectives*: Specific and measurable goals and objectives that outline the desired outcomes and milestones to be achieved along the journey of change *Action plans*: Detailed plans and initiatives that outline the specific actions, tasks and timelines needed to implement the strategy and drive the desired changes

3.3 Long-Term Goal-Aligned Alternatives

2. Apply the principle of **continuous learning with regular assessment of the impacts** on the areas where we are focusing the change and bearing in mind the dimensions of the triple transition. This continuous learning will allow for adjustments and improvements to ensure the sustainability and competitiveness of the innovations promoted over time;
3. **Continuous monitoring, evaluation and adaptation** of strategies and initiatives;
4. Recognise the **significance of circular economy principles**. This seeks to minimise waste, optimise resource utilisation and promote the reuse, recycling and regeneration of materials. By embracing circularity, we can reduce environmental footprint, enhance resource efficiency and create new economic opportunities through innovative business models and practices.
5. **Resilience** building adaptive capacity and robustness within the organisation or the ecosystem to withstand shocks and disruptions. This includes diversifying the economic base, fostering innovation and technological advancement and promoting a culture of continuous learning and adaptation. By enhancing resilience, we will be better equipped to navigate uncertainties, seize opportunities and ensure long-term economic stability and prosperity; and
6. **Anticipation** to initiate and stimulate more disruptive co-creation processes:

    - to start acting on time and move things forward on the right direction,
    - to transform inefficient structures and processes, and unfair and unequal dynamics into win–win visions and actions;
    - to gain momentum to create a true culture of change where a human-centred and planet-friendly transition can take the driving wheel of our actions, of our companies and of our innovation ecosystems (Table 3.5).

Strong impact hypothesis in forecasting approach to anticipate problems in alignment with the institutional, local/regional, national and international agendas and be ready to give solutions when needed. If we do not have a clear mission

Table 3.5 Element 6 of the compass

Element of a compass		Eoh-for-good features
	6. *Mirror* Let's you see the compass face and distant objects at the same time (using the sight) which provides more accurate readings. The mirror can also double as an emergency signalling aid *Sight*: Can be helpful to aim more precisely at a distant landmark, especially in open terrain *Magnifier*: For more detailed reading of map features	*Balancing short-term gains with long-term sustainability* in the processes of transition refers to the conscious effort of organisations and innovation ecosystems to optimise immediate benefits while ensuring their actions align with long-term sustainable goals. It involves making strategic decisions that prioritise both short-term success and the preservation of resources, environmental stewardship and social responsibility for lasting positive impacts

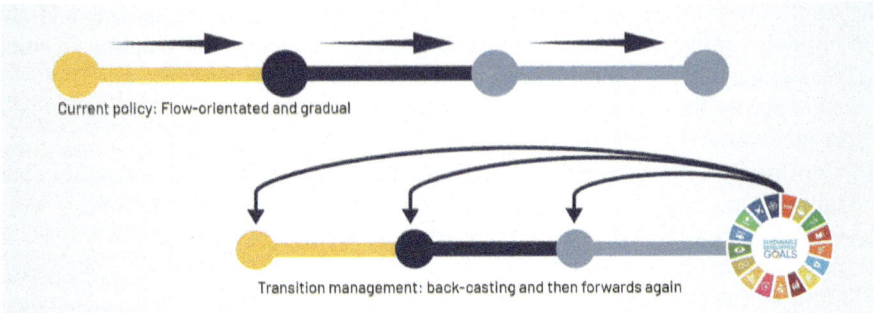

Fig. 3.2 Current policies and transition management

statement defined with forecasting and anticipation mechanisms, we will basically always be late in the decision-making process and in the delivery of the solutions to the problems identified (e.g. forward-looking research) (Fig. 3.2).

This means installing learning feedback and feedforward loops as we move forward. It requires a collective commitment that prioritises the common good and works towards shared outcomes and impacts that benefit all of humanity based on ethical long-term perspective decision-making to:

- make choices that promote social justice, environmental sustainability and inclusive economic development;
- prioritise the common good over short-term gains; and
- consider intergenerational equity and the preservation of natural resources.

The forward-looking vision to just socio-economic and environmentally friendly sustainability is underpinned by the principle of quadruple helix collaboration, which involves the active involvement of government, industry, academia and civil society in decision-making and problem-solving. This collaborative approach raises knowledge exchange, cross-sectoral partnerships and co-creation of innovative solutions that address complex economic and societal challenges. It recognises that diverse perspectives and expertise are essential for driving inclusive and sustainable economic development.

Living labs, as experimental environments where collaborators can co-create and test new ideas and solutions, play a crucial role in realising this forward-looking vision. They provide a space for open innovation, user-centric design and real-world experimentation, enabling the rapid prototyping and validation of innovative products, services and business models. Living labs promote collaboration, learning and the integration of multiple actors in the innovation process, ensuring that solutions are contextually relevant and address real-world needs.

As explained before, this forward-looking vision for the common good is deeply rooted in the principles of inclusion, access and agency. It recognises the importance of equal opportunities that entrust all members of society. This involves breaking down barriers to entry, promoting diversity and gender equality

Fig. 3.3 Amartya Sen, Martha Nussbaum and Henry Richardson,[13] hosting of the conference of the human development and capability association (HDCA) held in September 2015

and ensuring that marginalised communities have a voice and agency in shaping economic activities and decisions. By prioritising inclusion and access, ecosystems can harness the full potential of their human capital, drive social cohesion and nurture sustainable and equitable economic growth.

An example of forward-looking and ambitious plan is the 2030 Agenda for Sustainable Development Goals (SDGs)[14] which is a comprehensive and universal framework and roadmap that has been created to cover important issues to improve the population's quality of life, care for the planet and sustainable development. Therefore, all individuals play a fundamental role and we must explore the tools necessary to participate and contribute at the individual, group, institutional, community or ecosystem levels.

The SDGs serve as a rallying point for actors committed to change, providing a common language, vision and set of priorities. They offer a shared understanding of the pressing issues facing humanity and provide a roadmap for collective action. The goals and targets guide efforts to eradicate poverty, promote prosperity, protect the planet and ensure peace and justice for all.

To effectively address global challenges, the integration of the SDGs into innovation and institutional change processes is crucial. Innovation plays a vital role in developing sustainable solutions, while institutional change ensures that systems and structures support and enable the achievement of the SDGs.

---

[13] Professor Senior Researcher Scholar, Kennedy Institute of Ethics, Georgetown University.

[14] The SDGs consist of 17 interrelated goals and 169 targets that encompass social, economic and environmental dimensions.

Integrating the SDGs into innovation processes requires a deep understanding of the goals and targets and their relevance to specific contexts. It involves identifying how innovation can contribute to the achievement of each goal and leveraging technologies, business models and social innovations to drive progress.

During my Ph.D., I delved into the work of two renowned scholars, Amartya Sen and Martha Nussbaum, that have been decisive to the field of human development and the underlying principles of the United Nations.[15] Their approach, commonly known as the capability approach, highlights the importance of expanding people's freedoms and capabilities as the central goal of development. Sen and Nussbaum's human development approach provides a holistic perspective that aims to improve the well-being and quality of life for all individuals across the globe. By focusing on expanding people's freedoms and capabilities, promoting agency, inclusivity, fostering participatory decision-making and embracing environmental sustainability, this approach offers a valuable framework for understanding and addressing global challenges [10, 12] (Fig. 3.3).

Stay engaged as we proceed to Chap. 4, where we focus on establishing a culture of innovation and risk-taking to entrust individuals and organisations as agents of positive change. The journey continues to provide entrepreneurs in organisations and ecosystems with actionable insights and tools to shape a more resilient and adaptive future for our world.

References

1. Velasquez M, Andre C, Shanks TSJ, Meyer MJ (2018) The common good. *Issues in Ethics* 5(1). https://www.scu.edu/ethics/ethics-resources/ethical-decision-making/the-common-good/. Accessed 11 June 2023
2. Hussain W (2018) The common good. In: Zalta EN (ed.) The stanford encyclopedia of philosophy (Spring 2018). Metaphysics Research Lab, Stanford University. https://plato.stanford.edu/archives/spr2018/entries/common-good/. Accessed 11 June 2023
3. Goleman D, Boyatzis R, McKee A (2016) Primal leadership: unleashing the power of emotional intelligence. Harvard Business Review Press. Accessed 20 Jan 2023
4. Ackoff RL (1974) Redesigning the future: systems approach to societal problems. Wiley
5. Ackoff RL, Fred E (2005) On purposeful systems: an interdisciplinary analysis of individual and social behavior as a system of purposeful events (1st edn). Routledge
6. Ackoff RL, Gharajedaghi J (2010) Systems thinking for curious managers. Triarchy Press Ltd
7. Schein EH (2016) Organizational culture and leadership (5th edn). Wiley
8. Drucker P (2008) Management. Collins
9. Stuart E, Samman E (2017) (October 12). Defining leave no one behind. ODI: think change. https://odi.org/en/publications/defining-leave-no-one-behind/. Accessed 11 June 2023
10. Sen A (1999) Development as freedom. Anchor

---

[15] The UN 2030 Agenda is based on the Sustainable Development Goals.

References

11. Petrevska R, Caro-Gonzalez A, Bertello A, Bogers M (2023) Multi-vortex tornado blueprint for disruptive global co-creation (Inspired by EUvsVirus): Hackathons versus grand challenges. In: Facilitation in complexity: from creation to co-creation, from dreaming to co-dreaming, from evolution to co-evolution. Springer International Publishing. https://www.researchgate.net/publication/366979028_Multi-Vortex_Tornado_Blueprint_for_Disruptive_Global_Co-Creation_Inspired_by_EUvsVirus_Hackathons_vs_Grand_Challenges. Accessed 11 May 2023
12. Nussbaum MC (2013) Creating capabilities: the human development approach. Massachussets: Belknap Press: An Imprint of Harvard University Press

Open Access This chapter is licensed under the terms of the Creative Commons Attribution 4.0 International License (http://creativecommons.org/licenses/by/4.0/), which permits use, sharing, adaptation, distribution and reproduction in any medium or format, as long as you give appropriate credit to the original author(s) and the source, provide a link to the Creative Commons license and indicate if changes were made.

The images or other third party material in this chapter are included in the chapter's Creative Commons license, unless indicated otherwise in a credit line to the material. If material is not included in the chapter's Creative Commons license and your intended use is not permitted by statutory regulation or exceeds the permitted use, you will need to obtain permission directly from the copyright holder.

Chapter 4
Establishing a Culture of Innovation and Risk-Taking

Imagine if we could start preparing ourselves from a young age to cultivate a culture of preparedness and anticipation.

This chapter highlights the importance of creating just and inclusive dynamics within organisations. Entities of different kind can tap into a broader range of perspectives, experiences and talents, by promoting diversity and inclusivity, leading to more innovative solutions. It explores how organisations can promote a culture that values and respects all individuals, ensuring that every voice is heard and valued in the pursuit of transformative goals towards the common good.

The chapter aims to explore the multifaceted dimensions of entrepreneurship and how they contribute to promoting self-organisation readiness for just transformative change. Divided into three sections, it highlights key aspects of entrepreneurship from a systemic view. The sections encompass personal development and inner transformation, promoting autonomy, creativity and developing problem-solving skills. How to provide resources and support to overcome barriers and resistances to innovation.

Envision yourself motivated and trusted to put that idea in motion, with activities going as planned, there is positive impact beyond imagination and you are highly content with a job well done, network solidified and a nice future traced. Now imagine that this is step 1 to that scenario. Let's trace the path that helps us navigate as we go along.

How flexible are institutional governing instances and leadership to promote processes of change? Are they able to channel both intrinsic and extrinsic innovative persons and ideas and encourage continuous learning and personal and professional growth?

How to prepare to nurture and channel intra and entrepreneurship?

Establishing a culture of innovation and risk-taking within an organisation yields numerous benefits in the context of addressing the process of change. Firstly, such a culture promotes creativity and forward thinking, inspiring employees to generate novel ideas and approaches. This leads to the development of innovative products, services and solutions, thereby enhancing the organisation's competitive advantage in the market.

Secondly, raising a culture of innovation and risk-taking cultivates a mindset that embraces experimentation and learning from failures. By motivating employees to take calculated risks and view failures as valuable learning opportunities, organisations can foster a culture of continuous improvement and adaptability. This enables them to navigate through dynamic and unpredictable environments more effectively, as they can quickly respond and adapt to changing circumstances.

Through open and inclusive processes, organisations demonstrate their commitment to fairness, accountability and shared goals. Transparency builds trust among all interested parties, fostering a collaborative culture where individuals feel safe to take calculated risks and experiment with new ideas.

A culture of innovation and risk-taking enhances employee engagement and satisfaction. It entrusts individuals by providing them with the freedom to explore new ideas and make meaningful contributions, thereby increasing their sense of ownership and fulfilment in their work.

The question of 'Who does the things?' is crucial. Plans require individuals to put them into action and changes rely on people to embody them.

> Let's remember that people and their relationships are the core of any strategy.
> Motivating and mobilising committed innovators is essential for driving institutional or systemic change.

Strategic plans without a clear understanding of who will execute them often end up being too idealistic. In fact, many plans fail because of the mediocre efforts of those responsible for implementing them. Thus, the importance of talent development, attraction and retention and of nurturing a culture of innovation, creativity and intra-entrepreneurship. When institutions support and entrust individuals and provide opportunities for fulfilling careers, motivation and intra-entrepreneurship is triggered, and the chances to deliver more and better results are enhanced.

Nurturing innovation, intrapreneurship and agility within organisations tap into the creative potential of employees unleashing a rich pool of ideas and experiences. This culture of innovation and risk-taking becomes a catalyst for organisational change and improvement and it can be cultivated by:

(a) entrusting and supporting intrapreneurs which can lead to continuous improvement of processes, the development of new products and services and the drive of growth, competitiveness and value; and
(b) bringing together individuals from various backgrounds, expertise and perspectives. This diversity of thought cultivates an environment that values and encourages innovation.

When interested parties feel heard and entrusted to contribute, they are more likely to take risks, propose new ideas and challenge conventional thinking.

> Resistance to change and risk aversion can hinder the development of intra-entrepreneurship within organisations.

4.1 Fostering Intra-entrepreneurship for Individuals to Become Agents of Change

Creating a culture that values and encourages creativity is essential for individuals to unleash their innovative potential. This can be achieved by:

- promoting entrepreneurship, open communication, diverse perspectives and experimentation;
- motivating individuals to think outside the box;
- challenging conventional wisdom and
- generating new ideas.

Entrepreneurship goes beyond the traditional understanding of starting a business. It encompasses a mindset and approach that promotes innovation, creativity and taking initiative from the individual to the systems level. It involves:

- embracing uncertainty;
- leveraging opportunities and
- adapting to dynamic environments.

> By promoting entrepreneurship, we can challenge and rethink the fundamentals of how current systems and structures operate, paving the way for new paradigms and possibilities and a sense of creativity that stimulates innovation within the organisation.

Organisations can keep individuals motivated and channel their maximum potential along their personal and professional careers by nurturing an environment that values and nurtures each individuals' qualities.

4.2 The Role of Leadership in Cultivating Intra-entrepreneurship

Leadership plays a pivotal role in enabling entrepreneurship and self-organisation. Inclusive-driven and compassionate leaders can ensure that the benefits of systems innovation are accessible to all through the promotion of a shared language and the encouragement of active participation.

These leaders inspire and motivate others, provide guidance and support and a collaborative work culture. They create an environment where individuals are encouraged to take risks, think creatively and contribute their ideas towards system-level change.

Lead by example plays a crucial role for leaders who should demonstrate these qualities themselves. They inspire and motivate individuals to suit these qualities in their own personal and professional growth.

Constructive regular feedback and recognition is essential for maintaining motivation and channelling the potential of each actor. Individuals provided with insights into their strengths and areas for improvement are enabled to grow and develop further. Recognition of achievements and milestones, whether through formal rewards or informal appreciation, reinforces individuals' motivation and boosts their confidence in their abilities.

Providing employees with the **freedom and a certain level of autonomy to explore new ideas and solutions** is vital for cultivating intra-entrepreneurship. Organisations can promote autonomy by allowing individuals to have ownership and responsibility over their projects and tasks, granting them with:

(a) the necessary resources,
(b) decision-making authority,
(c) skill development,
(d) problem-solving capabilities and
(e) innovative thinking.

Providing clear goals and objectives, but at the same time, allowing flexibility in how these are achieved, help individuals exercise their autonomy and make decisions that align with their strengths and interests. This sense of own control entrusts individuals' motivation, as they feel valued within the organisation.

One key aspect of promoting entrepreneurship is **personal development and inner transformation**. Entrepreneurs that cultivate self-awareness are more prompt to explore their purpose and passions. This inner growth enables navigating challenges, managing ambiguity, developing resilience and staying motivated. It nurtures a sense of purpose that aligns with the larger goals (Fig. 4.1).

Fig. 4.1 Full moon night at Serignan Plage (own picture)

Practising meditation, self-reflection and mindfulness has been instrumental in my growth as a person and as a leader. Among many others, two books, "The Power of Now"; and "A New Earth" (Tolle 2004; 2006)[1], have supported my personal maturing process as a person and as a professional and leader. These books highlight the need for conscious awareness and the cultivation of presence in our daily lives. By taking the time to cultivate a calm and focused mind; I have:

- learned the importance of embracing the present and letting go of past regrets or future anxieties;
- developed inner stillness, which in turn has brought about clarity and insight in decision-making processes;
- nurtured a balanced and motivated inner state, which has had a positive impact on my ability to navigate challenges with resilience; and
- taught me the importance of being in tune with my own emotions and thoughts.

This self-awareness has not only helped me understand myself better but has also enhanced my ability to empathise with others.

As a leader, these skills have been invaluable in building strong relationships and fostering a positive and collaborative work environment. Furthermore, I have found that I am better equipped to handle stressful situations and make sound decisions.

---

[1] Tolle's concepts are rooted in personal development and inner transformation.

4.3 Creating Intra and Entrepreneurial Just and Inclusive Dynamics

Intrapreneurs are individuals who possess an entrepreneurial mindset and drive innovation within the corporate environment. However, they often face various challenges and roadblocks that can hinder their efforts.

Intrapreneurs need an **organisational culture that encourages risk-taking and experimentation**. This can be achieved by cultivating a safe environment where failures are viewed as learning opportunities rather than setbacks. Entrusting intrapreneurs to test and iterate their ideas without fear of punishment or negative consequences enables them to overcome resistance and push boundaries.

By nurturing a supportive and inclusive environment, entrepreneurs can be entrusted to lead innovations and teams to contribute their unique talents and perspectives. This is vital for driving innovative solutions and achieving organisational and ecosystem readiness.

> Leaders in charge of the processes of change need to create a vision for the future that considers the interests of all involved and concerned actors [3]. This entails:
>
> - promoting trust and multi-actor and multi-level relationships;
> - developing a strategy that approaches that vision;
> - enlisting the support of the critical centres of power to achieve it and
> - incentivising the people whose actions are essential to implement the strategy.

Organisations should invest in providing continuous learning and relevant development and training programmes, specific to entrepreneurship, innovation management and leadership workshops and resources that enable individuals to acquire new knowledge, develop critical thinking abilities, and enhance problem-solving techniques to overcome barriers, communicate their ideas effectively and drive change within the organisation. This equips them with the skills needed to tackle complex challenges.

Cultivating a growth mindset stresses the belief that skills and abilities can be developed through continuous learning and perseverance. This is essential for encouraging a culture that embraces challenges. Individuals are more likely to take risks, think creatively and persist in problem-solving endeavours when failures are seen as opportunities for learning and effort and resilience is valued.

Intrapreneurs can greatly benefit from the **guidance and mentorship of experienced individuals within the organisation**. Establishing mentorship programmes where seasoned entrepreneurs or executives provide guidance, advice and support can help intrapreneurs navigate challenges, gain insights and refine their ideas. Mentors can provide valuable feedback, open doors to networks and offer insights on navigating organisational politics.

4.3 Creating Intra and Entrepreneurial Just and Inclusive Dynamics

One key aspect for organisations, intrapreneurs and ecosystems is **investment and securing the necessary allocation of resources**, including financial, human and technological to facilitate the necessary changes:

- to develop and implement innovative ideas and
- to support the implementation of the strategy.

Organisations should provide mentorship and establish dedicated funding mechanisms, including financial (such as innovation funds or venture capital funds), human and technological resources, specifically allocated to support the implementation of the strategy and the development of intrapreneurial initiatives. This support will enable intrapreneurs to **overcome financial barriers, navigate challenges, pursue their ideas with confidence and facilitate the necessary changes**. Linked with this are two important issues:

1. The first one deals with a mindset shift to manage the availability or constraints of time and resources.
   In general, we are used to driving by short-term approaches (e.g. policies, projects, initiatives or products). In many cases, we still do not have the mechanisms or the capacity and resources to make long-term investments and robust monitoring beyond the project life cycle, the annual financial report, the four-year political cycle, etc.;
2. The second one is predictability. The question that raises is how can we predict the long-term impact of a policy, an initiative, a project, a product or a service? In fact, if we had to explain the uses or impacts in the long run of innovations (research results, products, services, etc.), we would be in real trouble. We can use examples such as Edison's experiments. The transformation of the phonograph from a simple dictation machine to a popular music playback device demonstrated how an invention can take on a life of its own and be adapted in ways that its original creator may not have anticipated. Edison's initial vision for the phonograph was limited to its practical applications, but its eventual impact on the music industry and entertainment culture far exceeded those expectations.

Incentivising: recognising and rewarding intra and entrepreneurial efforts is crucial to motivate and reinforce positive behaviour. Publicly acknowledging and celebrating the achievements of intrapreneurs creates a culture that values and supports their initiatives. Rewards can take various forms, including financial incentives, promotions or career development opportunities. Recognising and rewarding intrapreneurs not only motivates them but also sends a clear signal to the rest of the organisation that innovative thinking is valued and supported.

The first step is to define clear criteria for recognising and rewarding initiatives. These criteria should align with the organisation's goals, values and strategic priorities. An evaluation process should be established to assess the initiatives against the defined criteria. This process can involve a combination of quantitative and qualitative measures, such as financial impact, market potential, creativity, collaboration, sustainability and societal and environmental impacts. By establishing clear criteria and a transparent evaluation process, employees have a clear understanding of what is expected and how their initiatives will be assessed.

Financial incentives can also be an effective way to recognise and reward entrepreneurial initiatives. Linking financial rewards to the success of the initiative motivates employees to take risks and drive results. Examples of tangible financial incentives include performance-based bonuses, grants for innovation projects or seed funding for entrepreneurial ventures within the organisation, profit-sharing or equity stakes in the venture.

Non-financial recognition is equally important in acknowledging and rewarding entrepreneurial initiatives. This can take various forms, such as public recognition, certificates, awards or promotion opportunities. This can be channelled through publicly acknowledging employees' efforts through company-wide announcements, internal newsletters or dedicated events. These create a sense of pride and foster a culture of recognition. Other forms of non-financial reward are providing opportunities for professional development, mentorship or leadership roles within the organisation.

Innovation challenges and competitions are powerful recognition ways as these initiatives boost employees to think creatively and come up with innovative solutions to specific problems or opportunities. Examples include hackathons, idea pitches or innovation tournaments where individuals or teams compete to develop the most promising ideas. Winners can be rewarded with financial incentives, recognition or the opportunity to develop their ideas further within the organisation.

Training programmes, conferences, workshops or educational resources are also strong recognising opportunities, they build up on the employees' professional development, enhancing their skills to support their entrepreneurial aspirations. For example, organisations can sponsor employees to attend innovation-focused conferences, provide access to online courses on entrepreneurship or facilitate participation in industry events and networking opportunities.

> Recognising and rewarding entrepreneurial initiatives goes beyond incentives and recognition. It also involves creating a supportive environment that provides the necessary resources and infrastructure for individuals to pursue their entrepreneurial ideas. This can include dedicated innovation labs, access to mentors and experts, collaboration spaces and technological tools. A conducive environment enables individuals to take risks, experiment and drive their initiatives forward.

One example of recognising and rewarding entrepreneurial initiatives is Google's '20% time' policy, where employees are prompted to spend 20% of their working time on side projects and innovative ideas. This initiative has resulted in the development of successful products such as Gmail and Google Maps, providing employees with recognition and opportunities for career advancement.

Another example is the 'Innovation Awards' programme by International Business Machines Corporation (IBM), which recognises and rewards employees who have made significant contributions to innovation within the company. The programme includes financial rewards, public recognition and the opportunity to present their work to senior executives.

Building networks and encouraging collaboration are essential for intrapreneurs to leverage collective intelligence, collaborate and learn from each other. Establishing communities of practice, cross-functional teams, innovation labs, interdisciplinary and intersectoral platforms or innovation forums allows intrapreneurs to connect with like-minded colleagues, share knowledge, exchange ideas and leverage diverse perspectives. Collaboration can lead to the emergence of innovative solutions.

> When employees are confident that their contributions will be valued and supported, they are more likely to embrace innovation and drive organisational change towards the common good.

As we proceed to the next chapter, we will delve into the driving forces of system-wide transformative governance. We will explore collaborative elements and innovative co-creation vortices that start with an 'i', maximising the impact of multi-level and multi-agent governance processes. Our journey continues, encouraging entrepreneurs to become change agents in cultivating resilient and adaptive governance for the future.

References

1. Tolle E (2004) The power of now: a guide to spiritual enlightenment. New world library
2. Tolle E (2006) A new earth: awakening to your life's purpose. Penguin Books Ltd., United Kingdom, London
3. Medina AC (2000) Liderazgo y comunicación en la organización

Open Access This chapter is licensed under the terms of the Creative Commons Attribution 4.0 International License (http://creativecommons.org/licenses/by/4.0/), which permits use, sharing, adaptation, distribution and reproduction in any medium or format, as long as you give appropriate credit to the original author(s) and the source, provide a link to the Creative Commons license and indicate if changes were made.

The images or other third party material in this chapter are included in the chapter's Creative Commons license, unless indicated otherwise in a credit line to the material. If material is not included in the chapter's Creative Commons license and your intended use is not permitted by statutory regulation or exceeds the permitted use, you will need to obtain permission directly from the copyright holder.

Chapter 5
Driving Systemic and Multi-level Transformative Governance

> *...matchmaking, creating bonding, networking...*
> *Let's embark on transformative, holistic and impactful innovation journeys!*

This chapter explores how transformative governance can continually evolve and improve, ensuring that it remains responsive to the ever-changing challenges of the future. It underlines the value of continuous learning and innovation within multi-level and multi-agent governance processes. Organisations and innovation ecosystems can maximise their collective wisdom, leading to more effective and adaptive governance through knowledge sharing and cultivating a culture of openness.

> We aim to get the involvement of more partners working towards the common good, so that we, individuals, companies, entities and ecosystems can navigate innovation futures with a greater sense of credibility and a broader base of support.

Institutions and/or individuals[1] address issues differently, to different extents, lengths or depths. This depends on their nature, scope and features (large or small companies; public or private entities), vision, mission and strategic plans, interest and specific needs, resources, etc.

An enthralling highlight of this chapter is our exploration of the multidimensional co-creation vortices of transition—a collection of more than 20 dimensions, all starting with the letter 'i'. Through the creative interaction of these dimensions, we gain insight into the diverse facets of transformative change, shedding light on the intricate nature of transitions.

Let's start by presenting the needed collaborative dimensions and dynamics.

---

[1] SMEs, larger companies, universities, RTOs, public bodies, civil society organisations, foundations, and citizens, professionals, entrepreneurs, patients, etc.

5.1 An Overview of the Multi-i Collaborations for Transition—20+ Dimensions that Start with An 'i'

In this fascinating exploration, the following figure presents an overview of the diverse dimensions of transformative change, each denoted by an 'i'. These offer a holistic perspective, encompassing various facets of transition and fostering a deeper understanding of the complexities at hand.

The figure graphically illustrates a synthesis of the main elements and dynamics that we have been testing and experimenting in the last 20 years to promote change and governance processes in organisations, ecosystems and shared agendas. This will be explained along the chapter sections (Fig. 5.1).

These over 20+ elements that start with an 'i' intertwine as follows:

- 3 fundamental elements: innovation, inclusion and impact transversally tackled along the book as they underlie and guide the transition paths and the processes of change;
- 10 collaborative 'i's: interpersonal, intercultural, intersectional, intergenerational, interdisciplinary, interhelix, intersectoral, intra and interinstitutional, interregional and international;
- 9 'i's that represent the different phases of the process of change: from the initial idea to the ideation, intuiting, inspiring, integrating, interpreting and institutionalising. With investing and incentivising as important dimensions to reward and boost innovation and entrepreneurship.

> How does it work?
> Ad hoc in-house and in-context co-creative processes are designed and triggered by activating different (up to 20+) interrelated dimensions that start with an 'i', hence its 'multi-i' collaborative nature: interpersonal, interinstitutional, interdisciplinary.

For a systemic process of change in organisations and ecosystems to occur, several elements need to be integrated in a cohesive and interconnected manner.

1. The **effective deployment of multi-level innovative governance implementation and deployment dynamics and methodologies** which guide the collaborative process and nurture the practical realisation of innovative solutions.
2. The **grounding principles and shared visions** explained in Chap. 3. This derives from the combination of **collaborative efforts** 10 Collaborative i's through which individuals exchange ideas, share experiences and work together on projects. These enable interaction and co-creation within organisations and ecosystems and create opportunities for collective problem-solving, enhancing creativity and allowing individuals to learn from one another and leverage collective intelligence. Engagement should go beyond information

5.1 An Overview of the Multi-i Collaborations …

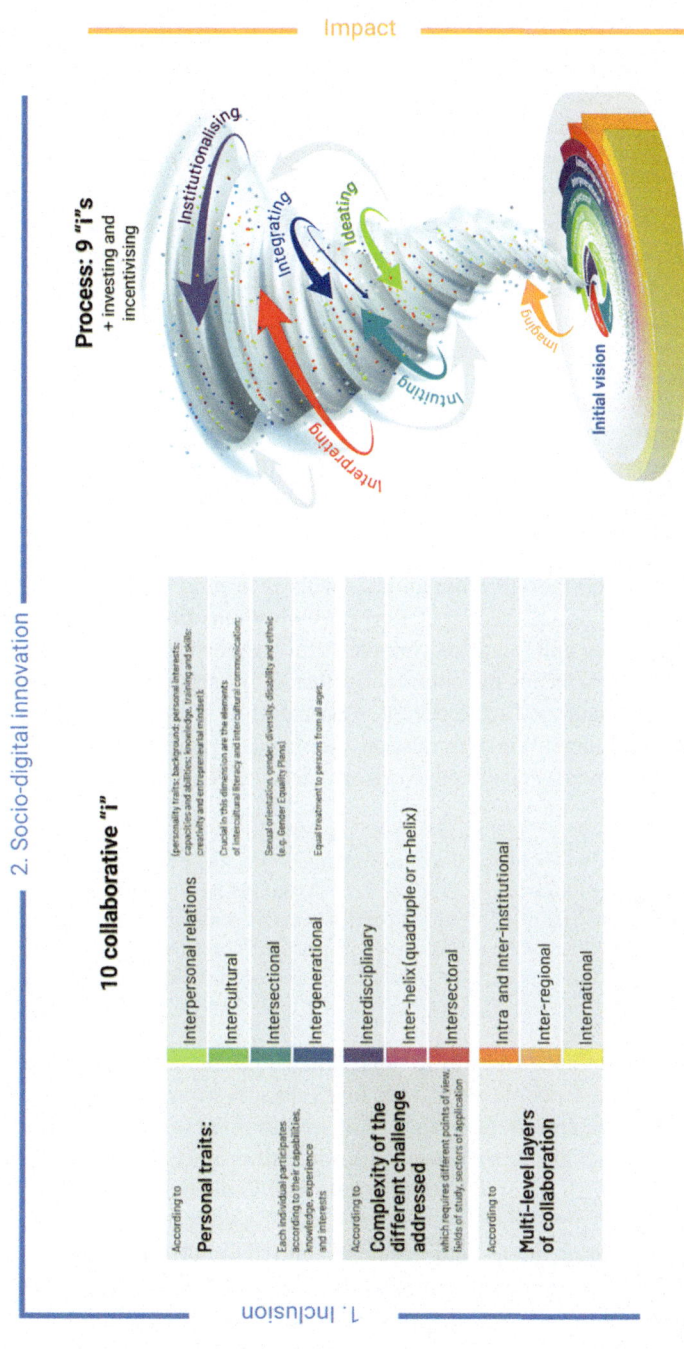

Fig. 5.1 Overview of the 20+ 'i' dimensions

dissemination and consultation, striving for active participation, entrusting actors and fostering lifelong learning and capacity building.
3. The **socio-digital innovations** that provide the socio-technical backbone for the multi-level governance process of change. These refer to the digital platforms, technological systems and networks that provide the necessary foundation for innovation and collaboration. Collaboration and co-creation platforms facilitate dialogue, coordination and joint decision-making, enabling collaborators to align their efforts and amplify their impact.

They enable the seamless exchange of information, resources and ideas among different actors, facilitating co-creation, knowledge sharing, problem-solving the investment and adequate allocation of resources which secure sustainability and boost innovative ideas to thrive.

5.2 The Deployment of Multi-level Innovative Governance Dynamics—The 10 Collaborative Elements that Start with An 'i'

The Eoh-for-Good collaborative dynamics revolves around 10 collaborative elements that start with an 'i', such as interpersonal connections, intercultural dialogue, interdisciplinary collaboration, intersectoral partnerships, etc.

Multi-i collaboration means the coming together of different people, organisations and actors to understand and solve complex problems in a more collaborative and mutually beneficial manner. Win–win relationships bring together diverse perspectives, knowledge and skills increasing effectiveness, efficiency, quality and sustainability.

> Collaboration promotes inclusive decision-making, collective ownership and shared responsibility leading to fairer outcomes and positive impacts. Through collaboration and co-creation among diverse interested parties we can establish a more comprehensive and negotiated collective vision by involving all relevant actors. This ensures that the deployment process considers the perspectives and needs of all actors involved.

Actively involving and consulting with interested parties, affected communities and marginalised people or at risk of exclusion from the outset is crucial for **meaningful engagement of actors fair deployment and positive impacts throughout the innovation process. Collaboration and knowledge sharing among individuals with diverse backgrounds and expertise** can lead to innovative problem-solving approaches.

It involves translating ideas and concepts into tangible actions, products or services that can create potential real-world positive impact. Implementation and deployment require new collaborative governing mechanisms to align shared goals and strategies, coordinated efforts and resource allocation across the different collaborators in the transformative innovation path

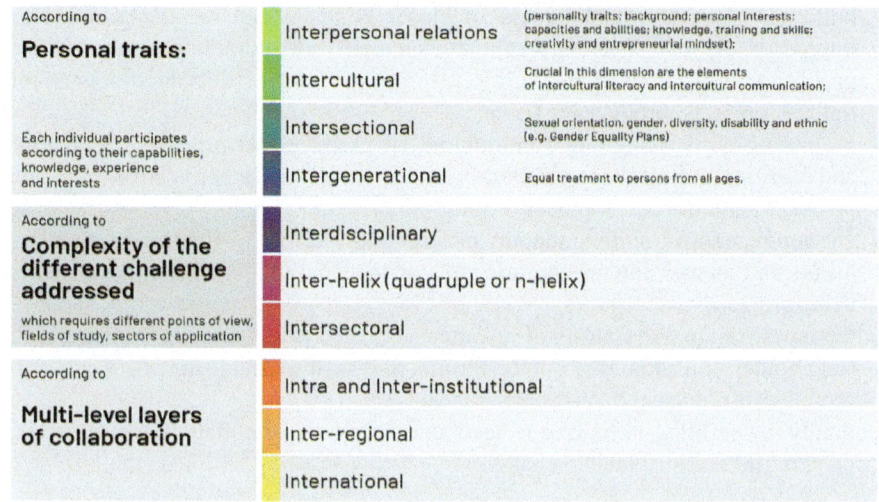

Fig. 5.2 The 10 collaborative 'i's

Drawing on a collective intelligence, that aligns and negotiates different needs, interests, rhythms, etc., they can lead to:

- transformative change that benefits the common good,
- enhances coordination, problem-solving capabilities and innovation,
- reduction of duplication of effort, and
- a more unified voice.

When interested parties collaborate and share their expertise, background, best practices and lessons learned. They generate novel approaches to complex challenges.

Effective collaboration is more likely to occur when participants have a clearly stated purpose based on shared values and interests. It is essential to recognise that the partners involved often value different things. The goal of collaborative work is to find common ground and generate collective action to improve the quality of the work. With an increased capacity to solve complex societal problems, multi-i collaborations can lead to significant and sustainable change.[2]

By consciously integrating these elements, combined in different ways and depths into the innovation processes within institutions and ecosystems actors committed to change can create synergistic and transformative governance dynamics.

These ten collaborative 'i' dimensions can be sorted out according to three attributes: (a) personal traits, (b) points of view, fields of study and sectors of application and (c) multi-layers of collaboration, as portrayed in the table (Fig. 5.2):

These co-operations facilitate:

---

[2] Definition provided by the Arctic Institute of Community-Based Research (AICBR).

- building trust and establishment of closer bonds which are the grounds for interpersonal, interdepartamental, intersectoral, collaborations, exchange and proliferation of knowledge, ideas, points of view, interests, methodologies, initiatives, projects, products or services;
- promoting networking and communities of practice that provide a collective and learning platform for problem-solving, enable the identification of shared priorities, and the development of joint initiatives;
- entrepreneurship training, support programmes, access to funding, networking events and mentorship programmes that address the unique challenges faced by these groups;
- the mobilisation and sharing of resources and efforts to achieve common goals;
- stakeholder consultations, public forums and/or the establishment of advisory groups that represent diverse interests and
- finally, assembling initiatives to leverage the power of collaboration in driving meaningful and sustainable change.

Diverse perspectives and expertise are essential for generating innovative ideas, ensuring accountability and fostering ownership of the transformation agenda.

These are briefly explained below:

According to Personal Traits

Interpersonal

An interpersonal relationship [1] refers to a strong partnership between people working on the same project or organisation. Partners working together must share a special bond to deliver their best. People must be honest with each other for a healthy interpersonal dynamic and, ultimately, a just positive workplace environment. In an organisation: on the one hand, non-existent professional friendship has negative impacts because it would create a one-brain decision situation, invite monotony and lack of collaboration would reduce performance. On the other hand, if there is a friendly atmosphere, it would create a direct impact on the organisational culture.

Interpersonal intelligence [2] is the ability to relate well to others and manage relationships. It can be seen as the ability to communicate and understand and interact effectively with others. It can also be seen as the ability to recognise distinctions between other people, react appropriately to their needs, understand their actions and feelings and appreciate such perspectives with empathy.

Interpersonal relations happen when we generate relationships of trust, honesty and respect among people. These are crucial to build and maintain interpersonal relationships in any given context or endeavour. When in place, their positive effects have a supportive impact on the atmosphere and morale of individuals. Interpersonal relations[3] and bonding between individuals help raise mutual

---

[3] Available at https://retos-operaciones-logistica.eae.es/las-habilidades-interpersonales-y-la-seleccion-de-personal/.

knowledge, build trust, higher levels of engagement and a self or group-motivating performance drive. Many differences can arise in a collaborative environment, so maintaining healthy relationships and fluid communication between colleagues is vital to a group's functioning.

Elements such as empathy,[4] affinity[5] and commitment[6] must be embraced in organisations that aim to have good results from their work teams to be able to enhance social energy.

Intercultural

Intercultural relations happen when we interact and collaborate with people from different cultures and cultural diversity comes into play.

Crucial in this dimension are the elements of intercultural literacy and intercultural communication. Several authors have attempted to define intercultural communication which refers to communication between people of different cultures [3: 28]. Intercultural communication is a symbolic, interpretative, transactional and contextual process in which people from different cultures create shared meanings [4: 46]. Intercultural communication refers to the effects on communicative behaviour when different cultures interact together.

Coaching services allow sharing concerns and problems and help resolve the cultural dilemmas[7] that arise in the collaboration.

This collaboration happens when people from different cultural backgrounds come together, adapting the rules of operation and the ways of working together. Depending on the cultural background, there are different modus operandi, based on subtle differences [5] going beyond superficial behaviours to understand the various underlying assumptions.

Leaders appreciate other points of view by delving into the different layers of culture and identifying the most important dimensions of the cultural frameworks involved, they can broaden their perspective capitalising on differences and enhancing enrichment. Intercultural collaboration aims to establish and sustain intercultural dialogues and two-way relations of mutual appreciation.

Overcoming differences can lead to business success and joint ways forward. Honest and respectful conversations and forms of collaboration of mutual interest are based on recognising the diversity of contexts and practices [6].

---

[4] Empathy is the cognitive ability to perceive another individual's feelings in an everyday context. It is also a feeling of affective involvement of one person that affects another.

[5] Affinity alludes to the attraction or suitability of characters, opinions, tastes or other circumstances between two or more people.

[6] Commitment refers to responsibility. A committed worker is a person who is involved in his work and when given the opportunity, acts in a way that goes beyond their own interest.

[7] https://interculturalcollaborations.com/. Accessed on 20 June 2023.

Intersectional: Leveraging diversity and inclusivity

Inclusive dynamics provide equal opportunities for all individuals, irrespective of gender, race or socio-economic background. Diversity and inclusivity are essential for nurturing entrepreneurship and building more robust self-organised emerging systems. By embracing diverse perspectives, experiences and backgrounds, entrepreneurs can unlock a wealth of innovative ideas and approaches to contribute to and benefit from systemic innovation. This requires breaking down barriers and promoting collaboration among different key players, including government, academia, businesses and civil society.

The Oxford Dictionary defines intersectionality[8] as 'the interconnected nature of social categorisations such as race, class and gender, regarded as creating overlapping and interdependent systems of discrimination or disadvantage' [8, 9]. Intersectionality is the acknowledgement that everyone has their own unique experiences of discrimination and oppression and we must consider everything and anything that can exclude people—gender, race, class, sexual orientation, physical ability, etc. This is invaluable for understanding why equity and inclusion must be prioritised.

We all have multiple and diverse identities that cannot be disentangled from each other: they all come together to define who we are and how we understand and interpret the world.

Bringing different visions to a socio-digital innovation process and/or partnership is invaluable for an ethical and egalitarian starting point. SDG 5 related targets advocate for equality for women and girls so that they have the same opportunities and guarantees for personal and professional development at all stages of their lives as men, breaking down structures that downturn them.

The European Commission published the report Approaches to include gender equality in research and innovation (R&I) to enable all talents to reach their fullest potential so they can contribute to ground-breaking R&I, regardless of their personal or social background. This is underlined in the EU Pact for R&I, which sets gender equality and equal opportunities for all as one of its core principles. To promote diversity in R&I and open its gender policy to intersections with other social characteristics, such as ethnicity, disability and sexual orientation.[9] The EC supports the development of inclusive gender equality plans (GEPs) and policies

---

[8] First coined by law professor and civil rights activist, Kimberlé Crenshaw back in 1989, intersectionality was added to the Oxford Dictionary in 2015 with its importance increasingly being recognised in the world of women's rights [7]. The term was developed to articulate how black women experienced sexism differently from white women and racism from black men.

[9] Article 2 of the Universal Declaration of Human Rights: Everyone is entitled to all the rights and freedoms outlined in this Declaration, without distinction of any kind, such as race, colour, sex, language, religion, political or another opinion, national or social origin, property, birth or another status. Available at https://www.un.org/en/about-us/universal-declaration-of-human-rights.

5.2 The Deployment of Multi-level Innovative Governance Dynamics ...

in line with the 2020 Communication on the new European Research Area (ERA) and ERA Policy Agenda 2022–2024

> Intersectional coalition building is critical to the success of diversity and inclusion efforts. Mutual support will allow for broader reach, wider impact and stronger relationships that can raise future opportunities for innovation and partnership.

Some of the potential benefits of intersectional collaboration are:

- boosting higher impact with fewer resources and a more efficient use of time by pooling resources and supporting each other's initiatives, greater impact is achieved;
- increased engagement and visibility through the collaboration across groups which amplifies the reach of initiatives and the exposure to learn more and engage;
- easier to gain community buy-in by increasing diverse representation, participation and engagement, highlighting the collaboration between groups and communities' throughout the innovation life cycle and
- inter and cross-group alignment and solidarity.

To properly address intersectional collaborations, some examples of indicators are

- Gender Equality Index and Diversity: https://eige.europa.eu/gender-equality-index/2022
- Non-discrimination/Inclusion: High Level Group on Non-Discrimination, Equality and Diversity.

 **Intergenerational**
This dimension is about treating everyone fairly, regardless of their age. It means avoiding age discrimination, whether someone is young, older or anywhere in between. The aim is to promote equality for people of all ages. Intergenerational relationships [10] typically focus on sharing knowledge, cultural norms and traditions, reciprocal care, support and resource sharing.

Current socio-economic and demographic trends are challenging intergenerational relationships.[10] Intergenerational approaches have the potential to address many social development priorities, such as the building of active communities (e.g., revitalising public infrastructure through community-based projects), promoting responsible citizenship, as well as addressing inequality and social exclusion.

---

[10] The 2009 UN International Experts Group meeting on 'Family Policy in a Changing World: Promoting Social Protection and Intergenerational Solidarity' recommended building partnerships and cooperation between youth and older persons organisations [11].

Fundamentally, organisations that take the time to understand intergenerational dynamics can develop a solid plan to pass the baton smoothly, from generation to generation, without losing all the knowledge, wisdom and experience accumulated over the years, and combined with the inputs, skills and better adaptation to the technologies of the new generations.

However, this intergenerational collaboration in which all actors can express their voice and agency [12] would not be possible with an isolated and compartmentalised approach. Instead, these efforts can only be achieved through collaborative, inclusive processes and concrete actions.

To understand how important, it is to foster intergenerational collaboration, we must refer to ageism, a concept described by the United Nations as stereotyping (how we think), prejudice (how we feel) and discrimination (how we act) towards others or oneself based on age. The latest United Nations report[11] on ageism [13] shows that one in two people in the world has ageist attitudes, which impoverishes the physical and mental health of older adults and reduces their quality of life, costing society billions each year with endless consequences for the people concerned and society at large.

Intergenerational collaboration[12] occurs when intergenerational projects or activities are purposefully planned to include one, or several aims and outcomes across the generations (e.g., all participants participating with equal treatment, respect and taking into consideration their needs, level of experience, skills, values and knowledge).[13] This collaboration can occur in a formal, non-formal or informal setting.

Promoting innovation through a participatory intergenerational process gives an active role to people from all ages to influence, adjust, test, uptake, etc. innovation in real time, by being part of the process.

According to Different Points of View, Fields of Study and Sectors of Application

Interdisciplinary

Interdisciplinary relations occur when we recognise that solving complex problems requires integrating methodologies, ideas, concepts and approaches from different disciplines. By engaging in dialogue and collaboration, we can approach information processing and idea generation from fresh perspectives, leading to innovative and impactful outcomes.

---

[11] Available at Global report on ageism.

[12] The European Network of Intergenerational Learning (ENIL) defines Intergenerational Learning as 'A learning partnership based on reciprocity and mutuality involving people of different ages where the generations work together to gain skills, values and knowledge' [14].

[13] An example is the role of Eurochild in nurturing collaborative processes that advocate for children's rights and well-being to be at the centre of policy-making. Children and youth participation entrust the young generation who is given agency and an opportunity to be heard.

5.2 The Deployment of Multi-level Innovative Governance Dynamics … 67

Despite 40 years of ongoing debate and the lack of clear definitions and terminology [15, 16], we can operationalise interdisciplinary as an activity that involves integrating knowledge, theories or methods from diverse fields, resulting in holistic and systemic solutions [17].

Interdisciplinary collaboration, as highlighted by [18], is a complex phenomenon where professionals from various fields come together to achieve common goals [19].

Multidisciplinary can be seen as a juxtaposition of disciplinary components, while transdisciplinarity involves integrating interdisciplinary approaches with the engagement of non-academic actors in the research process, addressing the 'applicability gap' [15, 20].

In fields like engineering, psychology, economics and biology, addressing complex problems such as global warming requires collaboration among diverse disciplines. Collaboration enables researchers to step out of their comfort zones, promoting personal and organisational growth.

Interdisciplinary collaboration brings fresh perspectives and sparks innovative thinking, leading to new research and innovation breakthroughs. Its application in projects raises valuable learning, incorporating unexplored viewpoints for enhanced outcomes and creative processes.

Recognising the significance of interdisciplinarity, universities are increasingly adopting it as a fundamental approach to optimise results [21].[14] This includes explicit commitments in strategic plans, sustained support for interdisciplinary education, recognition of interdisciplinary expertise and improved evaluation of interdisciplinary research and teaching.

By establishing mechanisms that encourage interdisciplinarity, especially for early-career researchers, we can unlock its potential for significant positive change.

The application of interdisciplinary collaboration in projects can yield valuable innovation paths, leading to enhanced outcomes by incorporating new and unexplored perspectives. Such collaboration generates synergies for creative processes, communication and paradigm shifts.

 **Interhelix (quadruple or n-helix)**
Quadruple helix collaboration encompasses the active involvement of academia, industry, government and civil society. It has emerged as a transformative approach to innovation and societal development recognising:

– that effective solutions to complex challenges require the collective intelligence, resources and perspectives of multiple collaborators; and

---

[14] Advocate for entrusting interdisciplinary research centres, creating opportunities for junior researchers and gaining recognition and support from national and European research funders to promote transdisciplinary commitment and impact.

– that innovation and societal progress require the synergistic efforts of multiple key players, each bringing their unique perspectives, knowledge and resources to the table.

The quadruple helix approach promotes a transdisciplinary and participatory approach to problem-solving, driving innovation through cross-sectoral collaboration. Quadruple or n-helix collaborative approaches enable interested parties to pool resources and expertise, exerting influence on policy-making processes and advocating for necessary reforms.

Within this dimension, living labs have gained prominence as collaborative spaces where key players come together to co-create, test, and implement innovative solutions. Living labs serve as real-world environments, allowing for the experimentation and validation of ideas in diverse contexts. They enable end-users to actively participate in the innovation process, ensuring that solutions are user-centric and address real needs.

The evolution of living labs has led to the emergence of a new generation characterised by a networked structure. Rather than operating as standalone entities, these living labs form interconnected networks, sharing knowledge, resources and best practices. This networked approach promotes cross-pollination of ideas and facilitates the scaling and dissemination of successful solutions.

The networked structure of the new generation of living labs aligns well with this vision, as it promotes collaboration and knowledge exchange across different regions and domains. It facilitates the creation of a vibrant ecosystem where interested parties can collectively work towards achieving the triple transition goals, sharing experiences and leveraging each other's strengths.

Intersectoral

The intersectoral relations occur when we see the need to resolve the issues that affect or need the coordination and collaboration of different sectors.

In cross-sector partnerships or intersectoral collaborations, institutions from different sectors[15] collaborate to address complex societal issues and promote a combination of socio-digital-business innovation. Intersectoral collaboration has two dimensions: horizontal and vertical [22].

---

[15] Sectors are used by economists to classify economic activity by grouping companies or clusters that operate in similar business activities. Some sectors are engaged in activities that involve the earliest stages of the production cycle, such as extracting raw materials. Other sectors involve the manufacturing of goods using those raw materials. Other companies are engaged in service activities.

First, the horizontal dimension links the mainstream sector with different sectors. This can be with other sectors, such as finance, justice, environment and education, and non-governmental representatives from the voluntary, non-profit and private sectors.

The second vertical dimension links different levels within a given sector. Collaboration along the vertical dimension helps to ensure better coordination and alignment of purpose between, for example, other groups of an organisation, such as frontline workers and centralised policy makers, various levels of governance (e.g. municipal and provincial counterparts, or different geographical regions of a country).

This is linked with the intra-institutional collaboration among colleagues within one unit, among departments or divisions, etc. Having in mind the common problem with a holistic view of the context in which to work helps the organisation to master its own 'piece of work'. This means not just focusing on the specific job you are working on but how it fits in with all your other projects and with the rest of the organisation's goals, dynamics and projects. To address this and to have an improved view of the bigger picture, you need to work with a cross-functional team with different experiences, rhythms, tools, points of view, etc. When the right circumstances are in place, this results in greater involvement in a more inclusive and motivating working environment.

> To resolve the problems, move forward and generate real changes or positive impacts in an area, sector or challenge in question, we need the combined work of different types of actors. If we take as an example the emergence created by the growing number of wildfires in different parts of the world, we will need from the most immediate intervention of the firefighters who must put out the fire that is already there, to the most remote of the care of the ecosystem forest with different uses, different care and in a long-term perspective.

According to The Multi-level Layers of Collaboration

 Interinstitutional
Institutional change goes beyond organisational boundaries and encompasses broader systemic transformations. It requires an understanding of the interconnections between institutions and their influence on various aspects of society, such as sustainable development, social equity and economic prosperity. By aligning institutions with these broader goals, it becomes possible to drive positive and inclusive societal change.

Institutions cannot work in isolation from its ecosystem. In the current hyper-connected world, each one has to seek a network of different types of organisations with which to establish meaningful partnerships to exchange and learn, support each other and find the balance between collaboration and competence. Establishing and maintaining relationships with other public and private, local, regional, national and/or international entities can enrich the work carried out and the products or services produced or delivered.

Interinstitutional collaborations can take various forms depending on the nature and objectives of the participating organisations. The following are just a few examples of the many types of interinstitutional collaborations that exist. The specific collaborations undertaken depend on the goals, capabilities and areas of interest of the participating institutions. These can be of different nature, thus linking with cross or intersectoral and international dimensions.

– **Research collaborations**: Institutions from academia, research centres and industry collaborating to conduct joint research projects, exchange and produce knowledge, share resources, create patents, spin offs, etc.
– **Consortia and partnerships**: Multiple institutions that come together to form a consortium or alliance, pooling their resources, expertise and infrastructure to work on common goals or projects.
– **Joint educational programmes**: Educational institutions collaborate to offer joint educational programmes, where students can earn a school, vocational training, university programme, etc., from two or more institutions upon successful completion.
– **Exchange programmes**: Institutions establish reciprocal exchange programmes that allow students, faculty or researchers to spend a period of time at partner institutions, promoting cultural exchange and academic collaboration.
– **Training and workshops**: Institutions collaborate to organise joint training programmes, workshops or seminars to enhance the skills and knowledge of their staff, faculty or students.
– **Technology transfer**: Academic institutions partner with industry or commercial entities to transfer their research findings or intellectual property for commercialisation or practical applications.
– **Policy development**: Institutions collaborate to develop policies, guidelines or standards in areas such as health care, environmental protection or governance, pooling their expertise and perspectives.
– **Shared facilities and resources**: Institutions share facilities, equipment or resources to reduce costs, improve efficiency and foster collaborative research or educational initiatives.
– **International collaborations**: Institutions from different countries collaborate on joint projects, research initiatives, student exchanges, or capacity-building programs to raise international cooperation and knowledge sharing.
– **Advocacy and lobbying**: Institutions with similar interests and goals form alliances or coalitions to advocate for policy changes, influence legislation, or address common concerns at local, national or international levels.
– **Artistic and cultural collaborations**: Institutions in the arts and cultural sectors collaborate on joint exhibitions, performances, festivals, or cultural exchange programs to promote artistic exchange and enrich cultural experiences.

- **Consortium purchasing**: Institutions join together to negotiate bulk purchasing agreements for goods and services, leveraging their collective buying power to achieve cost savings.
- **Data sharing and collaborative analysis**: Institutions share data sets, collaborate on data analysis, or establish data consortia to advance research, solve complex problems, or gain insights from diverse data sources.
- **Humanitarian and development initiatives**: Institutions collaborate to address global challenges, such as poverty alleviation, healthcare access, education or disaster response, by combining resources and expertise.

Interregional

When preparing organisational change, interregional cooperation is an important objective to consider as is central for addressing cross-border challenges and collaboratively developing the potential of different territories [23] (e.g. the EU cohesion policy). To support such cooperation, the European Regional Development Fund offers three main components: cross-border, transnational and interregional cooperation. These collaborations strengthen the institutional capacity of public authorities and actors, promoting efficient public administration through the exchange of expertise, good practices and experiences.

In the period 2021–2027, interregional collaboration will allow regions to use part of their allocation to fund joint projects anywhere in Europe. The new generation of interregional collaboration programmes (Interreg[16]) will facilitate Member States in overcoming cross-border obstacles and developing joint services. The Commission proposes a European Cross-Border Mechanism as a new instrument to harmonise legal frameworks for border regions and Member States, using the European Committee of the Regions[17] as a model for organising and implementing this partnership. This will offer regions and local authorities a formal say in EU legislation, meeting their needs effectively.

Closely connected to this type of collaboration is the Smart Specialisation Strategy or RIS3,[18] which highlights partnership and the pursuit of active involvement from various actors, including companies, research institutions, civil society organisations and national, regional and local authorities.

---

[16] https://ec.europa.eu/regional_policy/policy/cooperation/european-territorial/interregional_en.

[17] The European Committee of the Regions (CoR) is an EU advisory body of local and regional elected representatives from the 27 Member States. Through the CoR, they can share their views on EU legislation directly affecting regions and cities.

[18] Conceived within the European Commission's reformed cohesion policy, smart specialisation is a place-based approach characterised by the identification of strategic areas of intervention based both on the analysis of the strengths and potential of the economy and on an entrepreneurial discovery process (EDP) with broad stakeholder involvement. It is outward-looking and encompasses a broad view of innovation, including, among others, technology-driven approaches, supported by effective monitoring mechanisms.

> Understanding interregional collaboration involves recognising a verticality of actors, ranging from the most local to a broader scope within the region, with each having an equal role in the process, tailored to national and regional institutional structures.

An exemplary case is the 4-year H2020 project called URBANOME, which embodies quadruple helix collaboration to comprehensively address urban health and well-being. The project develops an integrative analytical framework for cities, identifying key health determinants. Through Urban Living Labs, policies and precision interventions are co-created and tested, involving citizens, industry, public authorities and academia, leveraging their experiences and networks. The pilots conducted in multiple cities explore environmental factors, susceptibility, gender differences and socio-economic disparities, culminating in evidence-based policy recommendations to mitigate urban health inequalities. URBANOME actively participates in the European Cluster on Urban Health alongside five other funded projects.

International

Internationalisation plays a vital role in our daily, professional or academic life, but also for commerce and trade, diplomacy, political and socio-economic stability, etc. In fact, international collaborations have emerged as a powerful force in driving change and cultivating innovation in organisations and innovation ecosystems in a hyper-globalised world. By bringing together diverse expertise, resources and perspectives from other regions or contexts in the world, these collaborations offer unique opportunities for transformative growth [24].

International collaborations **facilitate cross-cultural exchange**, enabling organisations to tap into a rich pool of knowledge, experiences and approaches. The interaction of diverse cultural perspectives encourages creativity, alternative problem-solving methods and challenges conventional thinking.

International joint efforts **provide organisations with access to global talent**, allowing them to leverage a diverse range of skills, expertise and perspectives. By engaging with experts from different countries and backgrounds, organisations can assemble high-performing teams with complementary capabilities. This diverse talent pool enhances the capacity for innovation, fosters knowledge sharing and stimulates the emergence of novel ideas.

International partnerships offer **access to shared resources and infrastructures**, allowing organisations to leverage existing capabilities and facilities in different regions. This includes research laboratories, testing facilities, funding mechanisms and intellectual property. Such access can accelerate research and development, reduce costs and enhance the scalability of innovative solutions. Collaborations also enable the pooling of resources to address grand challenges that require extensive investments and expertise.

Promoting 'glocal' synergies **open doors to connect the local and global spans. Although innovation is grounded in specific locations and requires**

local implementation, yet it is crucial to ensure that local strengths are **in harmony with common 'glocal' objectives** (e.g. EU priorities, SDGs, other regions' needs).

By partnering with organisations in different countries, organisations can tap into foreign market insights, access distribution channels and navigate regulatory frameworks. This expansion provides opportunities for scaling up innovations, reaching a broader customer base and driving economic growth, and open up **new markets that facilitate the commercialisation of innovative products and services**.

International coordination **nurtures collaborative learning and knowledge transfer** between organisations through joint research projects, academic partnerships and industry collaborations. Organisations can share best practices, exchange technological advancements and learn from each other's successes and failures. This collective learning nurtures a culture of continuous improvement, accelerates innovation cycles and builds long-term capabilities.

> Just and sustainable practices can enhance the overall reputation and attractiveness of the organisation or the innovation ecosystem, attracting talent, investment and collaboration opportunities. Investors and interested parties increasingly prioritise sustainable social and planet-driven considerations in their decision-making. And demonstrating these commitments are becoming more important to attract funding and support.[19]

By involving diverse perspectives, local knowledge and lived experiences, innovation processes can be tailored to specific contexts, addressing social inequalities and ensuring equitable outcomes.

5.3 Forming Multi-i Collaborations

This section explores the power of collaborative vortices in driving transformative governance. By bringing together diverse actors and stakeholders, co-creative multi-i tornado become hubs of innovation, generating transformative solutions to complex problems faced by entrepreneurs, teams, organisations and ecosystems.

We delve into real-world examples of successful vortices and how they have sparked positive change in various contexts. The figure below captures the three main moments of the process of generating collaborative transformative dynamics (Fig. 5.3):

1. forming multi-i collaborations;
2. navigating multi-level and multi-actor governance processes; and
3. maximising shared outcomes and impacts.

---

[19] We are starting to see more catalytic capital, a capital that is willing to give up a bit of profitability, at risk, by means of achieving social objectives (e.g. SpainNAB work).

Fig. 5.3 Multi-i co-creative governance tornado

5.3 Forming Multi-i Collaborations

5.3.1 How to Start Processes of Change

The process of change is based on the given circumstances of a context has many uncertainties. Normally, these processes are underway and are hardly seen as they start with timid innovations from below (bottom-up).

On most occasions we do not know where to start. What drives us is a sensed need, an intuition or an incipient, still vague, idea of what we need to achieve, but we do not really know where to go or how to get there. Neither how the journey will be.

> An initial vision triggers us to start.
> An initial identified need and a forward-looking vision boosts the elementary steps.
> The initiators are committed entrepreneurs who search for others (peers, colleagues, allies) who might have some interest in, and benefit from, the initial vision.

Usually, a motivated (intra)-entrepreneur has a vision or that sensed idea of why it is necessary to drive that change. To turn the idea into reality, we start crystallising it into attempts, concrete initiatives or projects co-created, with several agents. There begins a rotating, still conventional movement, more typical of project management, following the stages of co-design, implementation, monitoring and further evaluation of the idea. Continuous learning helps to flexibly adapt based on the lessons learned (Fig. 5.4).

These emerging developments lead to novel configurations and co-creations. At this stage, incremental changes occur on a small scale, gradually building up an impetus that will manifest in subsequent phases.

In this sense, when an organisation or an innovation ecosystem set in motion the right circumstances multi-actor and multi-level co-creative vortices are formed. However, if the necessary elements are not provided or the change processes are not properly organised, these incipient initiatives might dissipate or fade

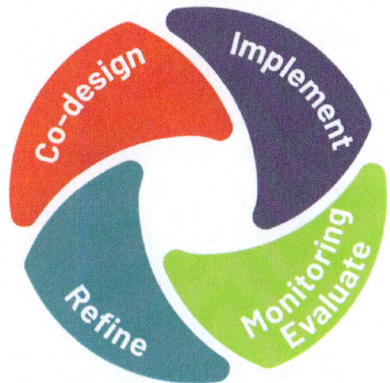

Fig. 5.4 Stages of the rotating movement

Fig. 5.5 Stages of the rotating movement and particles

away, like most tornadoes, which is the metaphor associated with this model. Just as a tornado requires a specific combination of weather conditions to form,[20] companies, entities, organisations need to prepare with anticipation for unforeseen circumstances to show resilience and readiness to adapt or quickly react to the given circumstances.

There needs to be a 'super cell'—a unique and precise alignment of factors—for the desired outcomes to materialise effectively, such as the promotion of an institutional culture for identifying and enhancing intra-entrepreneurship as explained in Chap. 3.

The base of the vortex rotates based on the combination of factors: the generation of new ideas, the capacity and willingness of the organisation to nurture intra and entrepreneurship, engagement of key players, right timing.

In organisations and innovation ecosystems, there are many particles (in the air, in the ground, at different altitudes) that are disconnected dots (Fig. 5.5).

In our visual way to explain the process, these are innovation particles (ideas, projects) that are activated by initiators, agents of change, leaders in different positions within an organisation or innovation ecosystem with vision and commitment to start processes of change. When a group of innovators and interested parties are identified and aligned, a whirling of co-design and co-creation movement in the base sets the process in motion.

---

[20] Strong contrasting cold-warm masses whirling horizontal and vertical winds…

5.3 Forming Multi-i Collaborations

> This is what we call the multi-i co-creative governance tornado for innovation. The active engagement, involvement and alignment of relevant interested parties, including internal and external partners, ensures collaboration and shared ownership of the process of change.

We have experienced that this active spiralling movement of co-design in the base, ignites different forms for collaborations: from interpersonal relationships to international competitive proposals that respond to different degree of needs (from community-based to grand challenges).

5.3.2 Navigating Multi-level and Multi-actor Innovative Governance Processes

Collaborative governance and anticipating management mechanisms and dynamics have increased exponentially thanks to the Internet and the socio-economic and technological progress achieved in a few decades (globalisation and interconnectedness, geopolitics).

We need better design and enhanced collaborative spaces for co-creation. These are physical or virtual environments specifically designed to raise collaboration, creativity and co-creation among diverse participants. Living labs, innovation hubs, co-working spaces and digital platforms serve as spaces where individuals from different departments, units or services within or across a company or an organisation (academia, industry, government and civil society entity) can come together to exchange knowledge, engage in experimentation and collectively develop innovative solutions to complex challenges. These collaborative spaces are the basis to boost the multi-i co-creative dynamics and vortices that are able to accelerate processes of change for the common good (Table 5.1 and Fig. 5.6).

The illustration shows a joystick that expresses the bottom-up, top-down and middle-round movement of the co-operations. It expresses the horizontal (combining different initiatives or innovations of 'i' collaborations) and the top-down, bottom-up, middle-round movement of the co-operations. These 360° strategic,

Table 5.1 Element 7 of the compass

Element of a compass		Eoh-for-good features
	7. *Declination scale* Used to orient the compass in an area with known declination and is also used for easily adding or subtracting the known declination in your area of travel	*Eoh-for-good encompassing management approach* triggers bottom-up, top-down and middle-round collaborative initiatives and processes within or across units, departments, divisions, institutions or ecosystems

Fig. 5.6 Ten collaborative 'i's with the movement of co-operations

Table 5.2 Element 8 of the compass

Elements of a compass		Eoh-for-good features
	8. *Rotating Bezel* Also called the 'azimuth ring', this outer circle has 360-degree markings. You hold the dial and rotate it to rotate the entire housing	*Multi-i co-creative whirlwinds* refer to specifically processes designed to facilitate collaboration, creativity and co-creation among diverse stakeholders. These processes materialise in spaces such as living labs, innovation hubs, co-working spaces and digital platforms that serve as platforms where individuals from academia, industry, government and civil society can come together

tactical and operational management can trigger initiatives and processes within or across units, departments, divisions, institutions or ecosystems.

Forming multi-i collaborations refer to processes explicitly designed to promote collaboration, foster creativity and encourage co-creation among a wide array of stakeholders (Table 5.2).

The base of the vortex rotates when the right circumstances are provided through ideas inception, intra and entrepreneurship, stakeholder engagement and continuous learning and adaptation. By exchanging knowledge, engaging in experimentation and collectively developing innovative solutions, these spaces act as the rotating bezel of a compass. These elements ensure alignment, collaboration, shared ownership, ongoing improvement and resilience in the face of change.

5.3 Forming Multi-i Collaborations

The funnel keeps rapidly growing by absorbing the innovation particles from the edges (interdisciplinary collaborations, local or international intersectoral projects, patents).

The initial intuition evolves as the process moves ahead. For keeping track of the change ahead, it is crucial the alignment of the leader or leaders with a continuous co-evolving forward looking vision for change (Fig. 5.7).

When the multi-i collaborations become stronger in organisations or innovation ecosystems, a more vigorous centripetal movement can suck more innovations into the co-creation funnel. This creates a whirlwind effect where innovation ideas are drawn in from different levels:

– the grassroots,
– the middle management, and
– the top leadership.

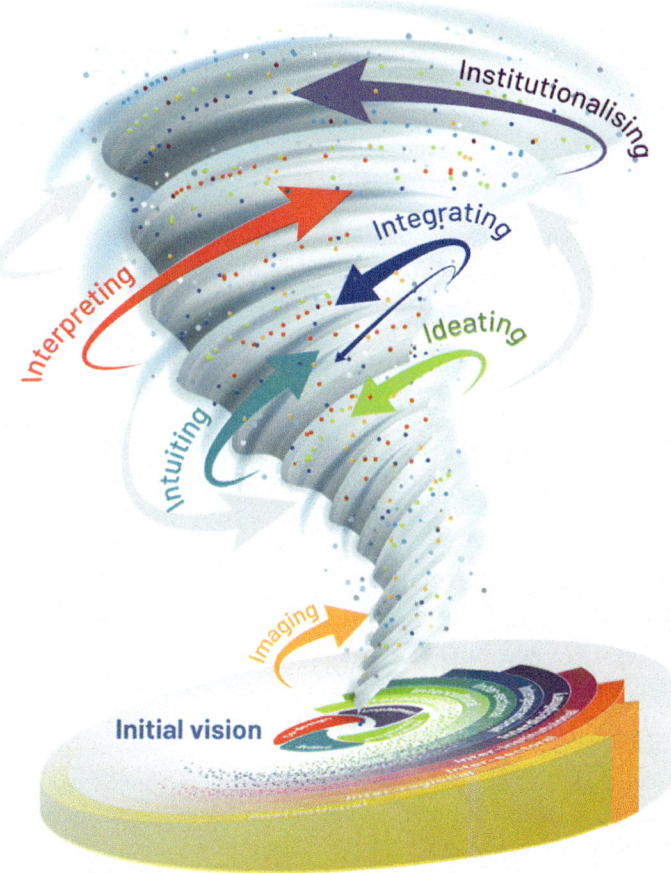

Fig. 5.7 Multi-'i' co-creative governance tornado

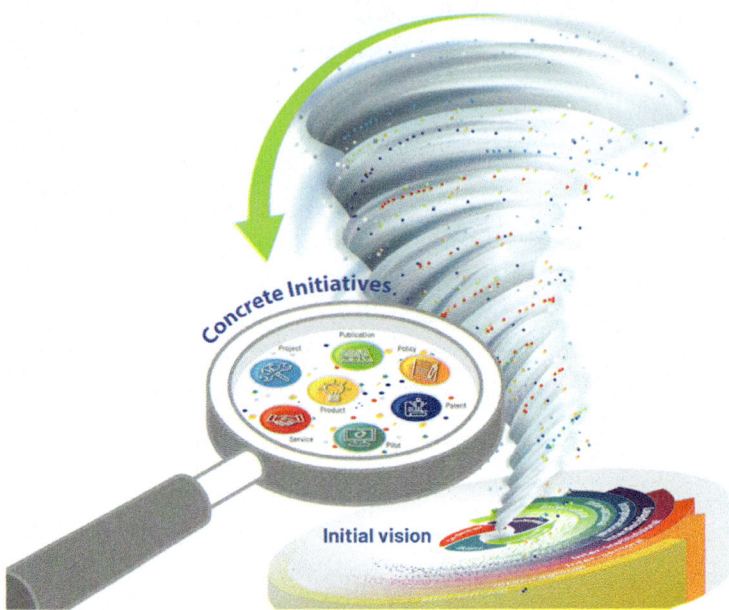

Fig. 5.8 Multi-'i' co-creative governance tornado and concrete initiatives

All the different dynamics of collaboration, initiatives, projects, concrete things are those particles that enter the co-creation vortex.

In the next phase of the model, we see that the spiral of the innovation vortex expands, attracting and sucking all these innovation particles from the edges (innovation cells attracted to the centripetal force of the vortex).

> Normally, innovation cells do not come from the mainstream or established system, but from the shores. That is where the multi-i co-creation mechanisms begin to gain relevance, when we see that all these innovative particles enter collaborative dynamics.

The way the process of change is materialised is through **concrete initiatives**, such as devising joint proposals, projects, patents, publications, adaptation of institutional processes to new regulatory frameworks, aligned with the shared vision, including the needs and interests of all parties involved (Fig. 5.8).

There are nine elements that comprise the different phases of the transition process. These are the 'i's in the centre of the illustration 19: From the incipient vision or idea that is quite intuitive and corresponds to the first stages: **ideating, intuiting, inspiring** then an innovation funnel grows based on the interactions and iteration of the different 'i's' during the process to interpreting data and the specific contextual circumstances (the givens of the context), integrating the elements and institutionalising initiatives.

5.3 Forming Multi-i Collaborations

Let's briefly explain the evolving stages in these multi-i innovation co-creative governance tornado.

1. From **intuition and ideation**, the base of the vortex rotates when the right circumstances are provided through ideas inception, intra and entrepreneurship, stakeholder engagement and continuous learning and adaptation, as explained before. Exchanging knowledge, engaging in experimentation and collectively developing innovative solutions act as the rotating bezel of a compass. These elements ensure alignment, collaboration, shared ownership, ongoing improvement and resilience in the face of change.
2. We **interpret** data on the current situation, what is to be achieved, the resources, the particles of innovation (who can we count on for, what is the real interest in the process of change). We interpret reality as we progress, we adapt based on two very important elements; flexibility and adaptability, (learning by doing) because we have to continually see what are the 'givens' and the evolutions of a specific situation, organisation or ecosystem.
3. **Integrate** the elements that we have (e.g. collaborative initiatives generated or being promoted, many times spontaneously, others with clear leadership). Here, we claim the need to boost professionalised accompanied processes (e.g. Eoh-for-Good expertise). This approach enhances synergistic dynamics fostering co-creation whirlwinds through conscious and better aligned decisions with a just triple transition. It empowers companies, organizations, entities and entrepreneurs to drive confident processes of change.
4. When the process of change stabilises, several initiatives get **institutionalised**, through investing, incentivising and aligning top-down, bottom-up and middle-round management mechanisms. This is the moment when plans or initiatives are promoted,[21] for example by creating incentives and rewarding mechanisms and by including specific measures in strategic plans, protocols and more. Clear examples are compulsory regulations, such as the plastic directive or the data processing directive which are mandatory for companies.

> When the entrepreneurial culture is nurtured and the conditions for change are triggered the impulse is boosted and the organisation can anticipate and position ahead.

The **stages of the change process are not sequential or linear**. The transformation is achieved through feedback and feedforward loops, going up or down in the vortex, applying the principle of continuous learning.

Constant adjustments are made as the process of change grows and scales with feedback and feedforward learning processes. These are resembled in the graph with the upward and downwards arrows.

We go some steps forward and then realise that some initiatives need further development or adjustments, certain pieces can better fit each other some other way and some initiatives may no longer respond to the expected outcomes and

---

[21] For example, the plastic directive, the data processing directive that are mandatory for companies.

should be discarded or abandoned. We learn through experimentation and through perseverance, continuous negotiation and persistent balance of needs and interests of actors and institutions transformed into win–win games.

5.3.3 Maximising the Learning and Innovation that Occurs in Multi-level and Multi-agent Collaborative Governance Processes

The outcomes of the initial vision flourish in the form of institutional transformation, shared agendas or quadruple or n-helix innovation ecosystems. The illustration above shows how collaborative dynamics can result in three different pathways (Fig. 5.9).

The first one is related to **institutional transformation**.

The second one is focused on the **definition of negotiated shared agendas**. Creating a shared vision and purpose provides a sense of direction and inspiration for change agents. The vision articulates the desired future of the institution or the innovation ecosystem (or a part of it) and the benefits that will be realised through the change process.

It is important that these agendas are not pre-designed in advance, but that all the interested agents are given a voice in the different initiatives, projects and actions where multi-i collaborations take place. This is so, for instance, in intergenerational collaborations, where listening to the different perspectives resulting from the dialogue between generations is crucial. Persons from diverse ages have very different ways of seeing and understanding life. Experts or professionals from different sectors can pose the problem from dissimilar perspectives. In most of the cases, solutions cannot be found without the support or vision of other sectors (e.g. understanding the validations that another company or laboratory has carried out on a given technology in other application domains).

The last one heads up towards more **cohesive innovation ecosystems** in which quadruple helix (or n-) helix participation is cultivated in the design, co-design,

Fig. 5.9 Three different pathways from collaborative dynamics

definition of challenges and governance processes. The outcomes and impacts of multi-level regenerative innovation extend beyond the individual organisations involved. Through knowledge spillovers, replication of successful models and the diffusion of innovations, the positive effects can ripple throughout the wider innovation ecosystem. This creates a virtuous cycle, where successful regenerative innovations inspire and catalyse further transformative changes in organisations, industries and society at large.

5.4 Reinforcing the Transition Gap—From Established to Emerging Systems

Heavy challenges, institutional tensions, market or revenue pressures, research and innovation readiness, etc., moisturising the soil for transition (Fig. 5.10).

During the initial phases or the first steps of an emerging system or process of change, innovations are introduced by socio-digital innovators and entrepreneurs, what I call, ***early adopters or agents of change***.

All these processes and co-creation dynamics and mechanisms begin with ideas, micro-projects and incipient co-creation initiatives (joining the dots) nurtured by a flexible and adaptable approach that create communities of practice and networks or constellations of innovators and innovations. At the beginning there is more like an artisan's job of uniting people and ideas into concrete initiatives (Fig. 5.11).

> Better trained agents of change could:
>
> - identify innovators, intra and interentrepreneurs;
> - align individual interests with those of the organisation and the systemic vision;
> - connect the dots through concrete initiatives;
> - build collaborations and propel ideas at the basis, from the bottom-up; and
> - map out the system dynamics and recognising emergent properties with anticipation.

When we join the dots and a number of actors get involved in flexible and adaptable collaborative initiatives, solid and functional bases are established. These involve reuniting actors coming or representing the different 'i's into flexible and adaptable partnerships (e.g. including civil society organisations, businesses, government agencies and academia: the quadruple or n-helix) (Fig. 5.12).

These are the grounds for the emerging system. As the process of change goes along, the systemic vision is gradually being shared, refined and co-evolved with the contributions of the different agents involved. These common vision and commitment develop a driving transformative change at scale with collaboration

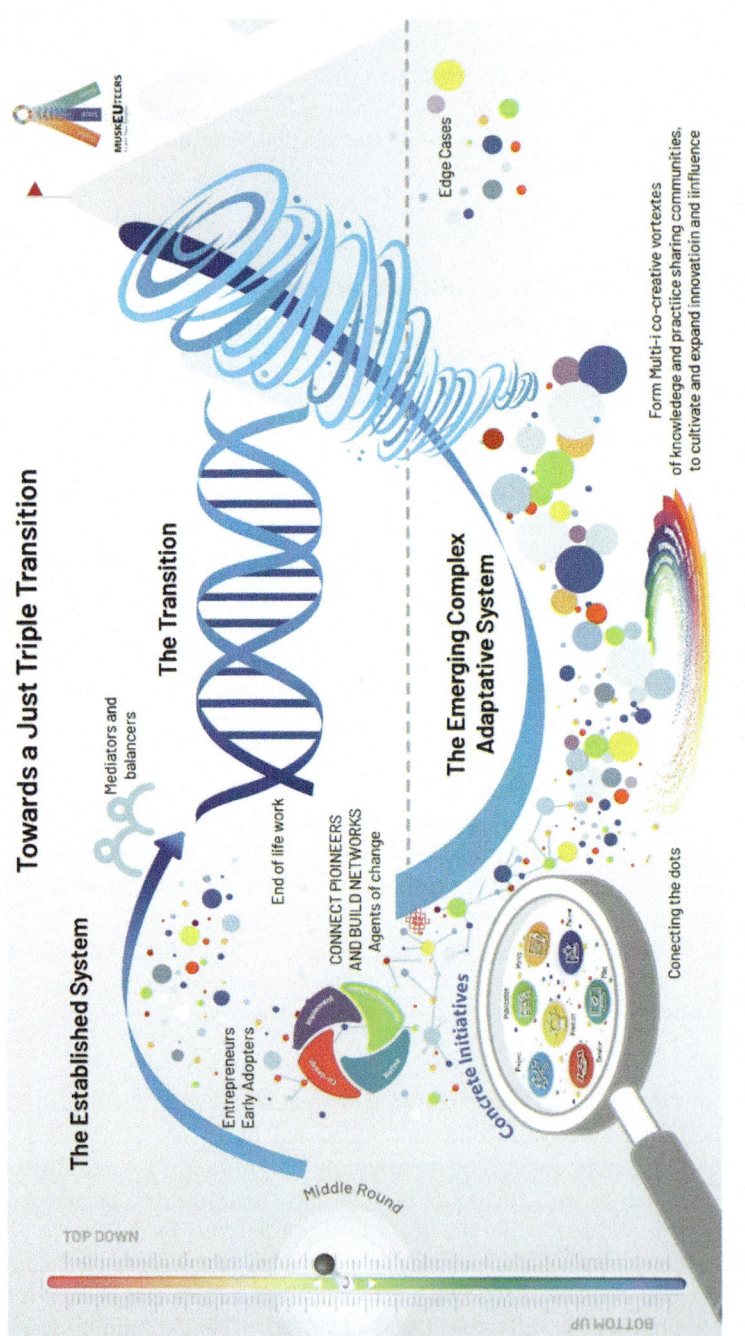

Fig. 5.10 Overview of the process of transition © Eoh-for-good

5.4 Reinforcing the Transition Gap …

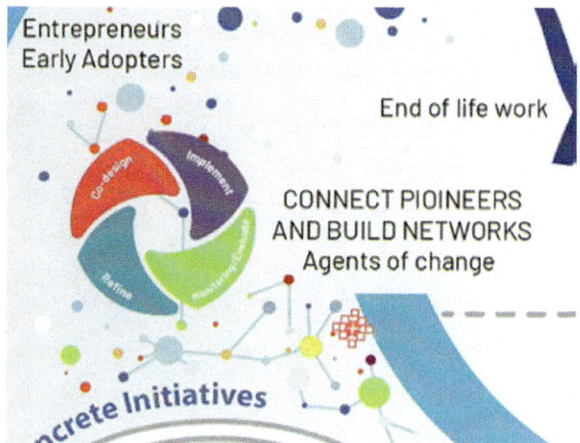

Fig. 5.11 Snapshot of the emerging phase © Eoh-for-good

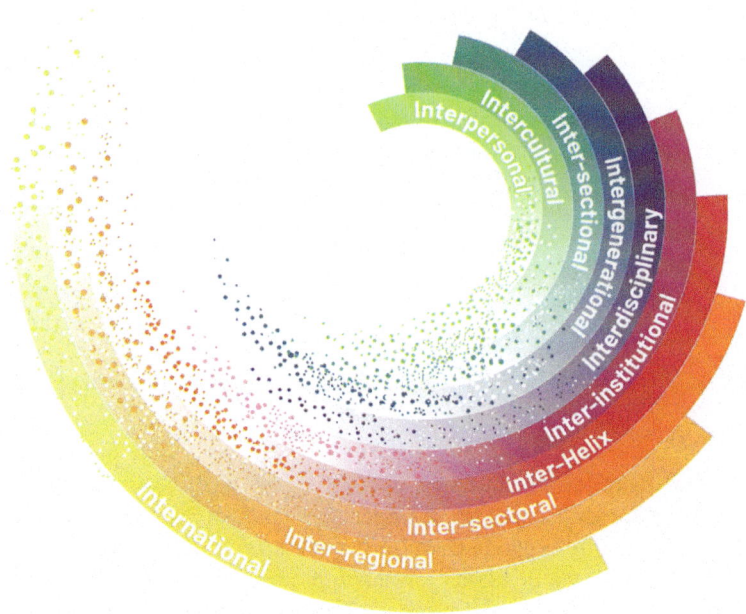

Fig. 5.12 Ten collaborative 'i's © Eoh-for-good

taking various forms, such as collaborative projects, multi-stakeholder partnerships, co-creation processes and participatory decision-making that put in place and strengthen multi-i innovation whirlwinds (Fig. 5.13).

Fig. 5.13 Snapshot of the spanning power of multi-i co-creative vortices © Eoh-for-good

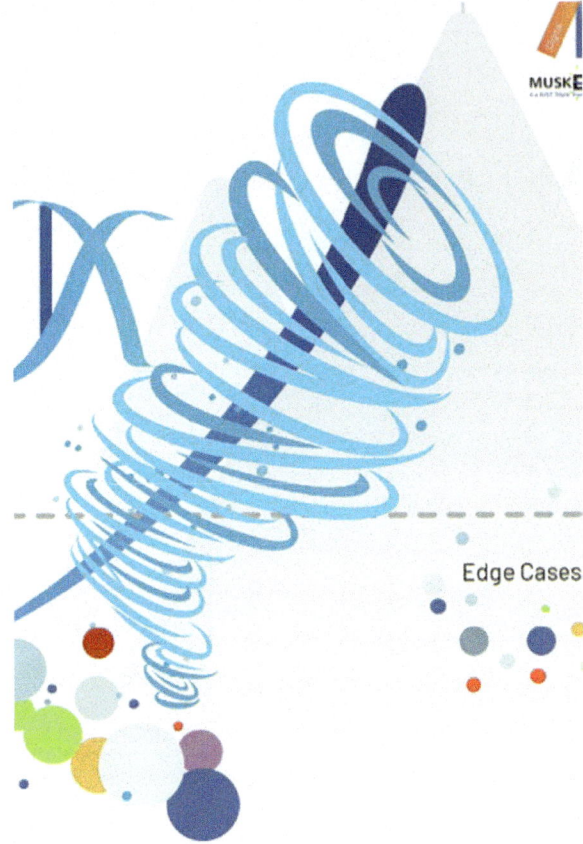

Form Multi-i co-creative vortextes
► of knowledege and practiice sharing communities,
to cultivate and expand innovatioin and iinfluence

When the organisation or the ecosystem enters a culture of innovation and risk-taking, the tornado becomes stronger and innovation is sucked from below and from the edges. It expands because it generates an ever-greater mass of collaborations that reflects in higher production levels and indicators (innovation, scientific indicators), increased number of projects with different types of collaborations that take place, etc. This is the moment when an organisation or an ecosystem begins to generate disruptive innovative co-creation dynamics and the emerging system is gaining momentum. This is usually accomplished by entrepreneurial processes capable of manoeuvring in a middle-round approach (middle management) (like in the joystick presented in the previous section) making the horizontal connections but also the convergence between the top-down and the bottom-up layers.

5.4 Reinforcing the Transition Gap ...

Fig. 5.14 Mediators and balancers

When we find ways to make these processes official or more integrated into the mainstream, the company, organisation, unit develop enhanced capabilities. A meeting point where ideas from the bottom-up (coming from individuals and grassroots initiatives) join forces with top-down regulation (established rules and guidelines) by the leadership to strengthen the existing dynamics. By doing so, initiatives and projects can be boosted to embrace these ideas and work towards a common goal that benefits everyone involved. It is all about finding the best way to negotiate and bring together different interests for a better outcome.

Depending on the type and nature of the organisation, the dynamics and targets will be different.

In this process, there are several individuals or entities that contribute to maintaining stability or equilibrium within the organisation or ecosystem during the process of change. We have called them: ***mediators and balancers*** (Fig. 5.14).

These are persons that intervene in conflicts or disputes to help find a middle ground or reach a mutually acceptable win–win resolution. They work to balance the interests and perspectives of different parties (units, departments, suppliers, external entities) involved in order to promote understanding and harmony.

'*End of care*' work refers to the compassionate and supportive efforts focused on managing and facilitating the end stages of a system's life cycle. It involves acknowledging that a system or model is no longer viable or effective and working towards its graceful and intentional closure. It involves providing care, guidance and resources to ease the process of letting go, ensuring a smooth transition to a new system or paradigm. It focuses on honouring the past while embracing the opportunities and transformations that lie ahead, promoting a nurturing environment for new developments.

> By working together, entrepreneurs (early adopters and agents of change) and key players can leverage their respective strengths and resources, leading to more effective, inclusive and just sustainable solutions. This collaborative approach also christalises in concrete initiatives that develop and promote a sense of ownership, shared responsibility and collective action, unity and solidarity. Multi-i and multi-level cooperation within the organisation develop more innovative integrated agendas, ecosystems, partnerships, networks and alliances committed to change.

References

1. Chaudhary AR (2019) Interpersonal relationship at workplace. All things talent. https://allthingstalent.org/interpersonal-relationship-at-workplace/2019/04/08/. Accessed 19 June 2023
2. Sadiku MNO, Musa SM (2021) Interpersonal intelligence. In: Sadiku MNO, Musa SM (eds) A primer on multiple intelligences. Cham: Springer International Publishing, pp 85–94. https://doi.org/10.1007/978-3-030-77584-1_7
3. Chen G-M, Starosta WJ (1998) A review of the concept of intercultural awareness. Hum Commun 2:27–54
4. Lustig MW, Koester J (2007) Intercultural competence. Allyn & Bacon. Accessed 13 June 2023
5. Ernawati DK, Sutiari NK, Astuti IW, Onishi H, Sunderland B (2022) Correlation between intercultural sensitivity and collaborative competencies amongst Indonesian healthcare professionals. J Interprof Educ Pract 29. https://doi.org/10.1016/j.xjep.2022.100538
6. Bird A, Osland JS (2005) Making sense of intercultural collaboration. Int Stud Manag Organ 35(4):115–132. https://doi.org/10.1080/00208825.2005.11043739
7. Crenshaw K (2017) On intersectionality: essential writings. Columbia Law School, The New Press, New York, NY. https://scholarship.law.columbia.edu/books/255
8. Perlman M (n.d.) The origin of the term intersectionality. Columbia journalism review. https://www.cjr.org/language_corner/intersectionality.php. Accessed 20 June 2023
9. Taylor B (24 Nov 2019) Intersectionality 101: what is it and why is it important? Womankind Worldwide. https://www.womankind.org.uk/intersectionality-101-what-is-it-and-why-is-it-important/. Accessed 20 June 2023
10. Hausknecht S, Neustaedter C, Kaufman D (2017) Blurring the lines of age: intergenerational collaboration in alternate reality games. In: Romero M, Sawchuk K, Blat J, Sayago S, Ouellet H (eds) Game-based learning across the lifespan: cross-generational and age-oriented topics. Springer International Publishing, Cham, pp 47–64. https://doi.org/10.1007/978-3-319-41797-4_4
11. United Nations Department of Economic and Social Affairs, Division for Social Policy and Development, Programme on the Family (2009) Report of the expert group meeting family policy in a changing world: promoting social protection and intergenerational solidarity. Doha, Qatar, p 37. https://www.un.org/esa/socdev/family/meetings/egmreportdoha09.pdf. Accessed 19 June 2023
12. Fonseca Peso J, Caro González A, Milosevic N (2020) Innovative co-creative participatory methodologies for a dreamt-of quality education in Europe. Sustainability 12(16):6385. https://doi.org/10.3390/su12166385
13. World Health Organization (WHO) (2021) Global report on ageism. Geneva. https://www.who.int/teams/social-determinants-of-health/demographic-change-and-healthy-ageing/combatting-ageism/global-report-on-ageism. Accessed 20 June 2023
14. ENIL (2011) ENIL—European network for intergenerational learning report on intergenerational learning and volunteering (p. 45). http://envejecimiento.csic.es/documentos/documentos/enil-ilv-01.pdf. Accessed 19 June 2023
15. Lyall C, Meagher L, Bandola-Gill J, Kettle A (2015) Interdisciplinary provision in higher education. Current and future challenges. https://www.researchgate.net/publication/303370206_Interdisciplinary_provision_in_higher_education_Current_and_future_challenges. Accessed 18 June 2023
16. Pohl C, Perrig-Chiello P, Butz B, Hadorn GH, Joye D (2011) Questions to evaluate inter- and transdisciplinary research proposals (Working paper). Swiss Academies of Arts and Sciences, td-net for Transdisciplinary Research, Berne. https://api.swiss-academies.ch/site/assets/files/14856/td-net_pohl_et_al_2011_questions_to_evaluate_inter-_and_transdisciplinary_research_proposals.pdf. Accessed 19 June 2023

References

17. Huutoniemi K, Klein JT, Bruun H, Hukkinen J (2010) Analyzing interdisciplinarity: typology and indicators. Res Policy 39(1):79–88. https://doi.org/10.1016/j.respol.2009.09.011
18. Mahdizadeh M, Heydari A, Moonaghi HK (2015) Clinical interdisciplinary collaboration models and frameworks from similarities to differences: a systematic review. Global J Health Sci 7(6):170–180. https://doi.org/10.5539/gjhs.v7n6p170
19. Houldin AD, Naylor MD, Haller DG (2004) Physician-nurse collaboration in research in the 21st century. J Clin Oncol 22(5):774–776. https://doi.org/10.1200/JCO.2004.08.188
20. Lawrence R, Despres C (2004) Futures of transdisciplinarity. Futures 36:397–405. https://doi.org/10.1016/j.futures.2003.10.005
21. Wernli D, Ohlmeyer J (2023) Implementing interdisciplinarity in research-intensive universities: (No. 3). Belgioum: League of European Research Universities, p 48. https://www.leru.org/files/Publications/Implementing-interdisciplinarity-in-research-intensive-universities-good-practices-and-challenges_Full-paper.pdf. Accessed 20 June 2023
22. Public Health Agency of Canada (20 April, 2016) Canadian best practices portal—CBPP. https://cbpp-pcpe.phac-aspc.gc.ca/population-health-approach-organizing-framework/key-element-6-collaborate-across-sectors-and-levels/. Accessed 20 June 2023
23. Uyarra E, Sörvik J, Midtkandal I (2014) Inter-regional collaboration in research and innovation strategies for smart specialisation (RIS3) (Technical Report No. 06). Joint Research Centre: European Commission, p 37. https://doi.org/10.2791/13682
24. Chen K, Zhang Y, Fu X (2019) International research collaboration: an emerging domain of innovation studies? Res Policy 48(1):149–168. https://doi.org/10.1016/j.respol.2018.08.005

Open Access This chapter is licensed under the terms of the Creative Commons Attribution 4.0 International License (http://creativecommons.org/licenses/by/4.0/), which permits use, sharing, adaptation, distribution and reproduction in any medium or format, as long as you give appropriate credit to the original author(s) and the source, provide a link to the Creative Commons license and indicate if changes were made.

The images or other third party material in this chapter are included in the chapter's Creative Commons license, unless indicated otherwise in a credit line to the material. If material is not included in the chapter's Creative Commons license and your intended use is not permitted by statutory regulation or exceeds the permitted use, you will need to obtain permission directly from the copyright holder.

Chapter 6
Designing Ad Hoc Impact Monitoring Systems

> *We can continuously improve and add value*
> *by meeting the evolving needs of communities and actors.*

This chapter explores the significance of evidence-based decision-making, providing valuable insights into creating comprehensive impact follow-up systems and a battery of indicators to assess the outcomes of transformative governance efforts. It highlights the importance of measuring progress and tracking the impact of transformative governance initiatives across various dimensions. By designing ad hoc impact monitoring systems, we can ensure that our efforts are aligned with the overarching goals of a just, sustainable and equitable future.

Institutions across sectors play a critical role in shaping societies, addressing complex challenges and producing societal, economic and environmental outcomes and impacts.

Up till now, Gross Domestic Product (GDP) has been the main measure of a country's economic performance. However, it is becoming clear that GDP does not capture the full picture of progress and well-being. The European Commission's Strategic Foresight Report [1] is taking a crucial step by exploring how to better measure progress and prosperity as they are developing Sustainable and Inclusive Well-being metrics that go beyond GDP. The goal is to include factors like quality of life, health, education, environmental impact and inequalities in measuring a country's well-being. The report introduces a pilot 'adjusted' GDP metric that considers the 'health' dimension of well-being, and further work will be done to incorporate other important factors.

In this new context, Environmental, Social and Governance criteria (ESG) have become a powerful framework that encourages responsible business practices, attracts investors seeking sustainable opportunities and promotes transparency and accountability. It is inspiring to witness the positive impact this holistic approach is having on the business world and our planet.

In a nutshell, the ESG framework can be described as the triple bottom line approach, encompassing environmental, social and governance factors.

The environmental dimension focuses on the effect of business activity on nature. Organisations must not only mitigate the possible adverse effects of their activity but are also required to undertake actions that generate a direct positive impact, for example, by reducing pollution, waste generation or the emission of greenhouse gases.

The social dimension relates with the communities where the company is present. It entails an analysis and an assessment of whether human resources promote equality and diversity among the workforce and social inclusion. Actions should seek to create a healthy space for human capital and the local community in general.

Governance encompasses all issues related to the company's corporate governance, purpose, culture, production processes and management. Corporate governance includes the following tasks: consider the composition and diversity of the board of directors, develop a code of ethics and best practice guide, check the supply chain to ensure compliance and provide transparent fiscal information in its accounts and all public reporting (Fig. 6.1).

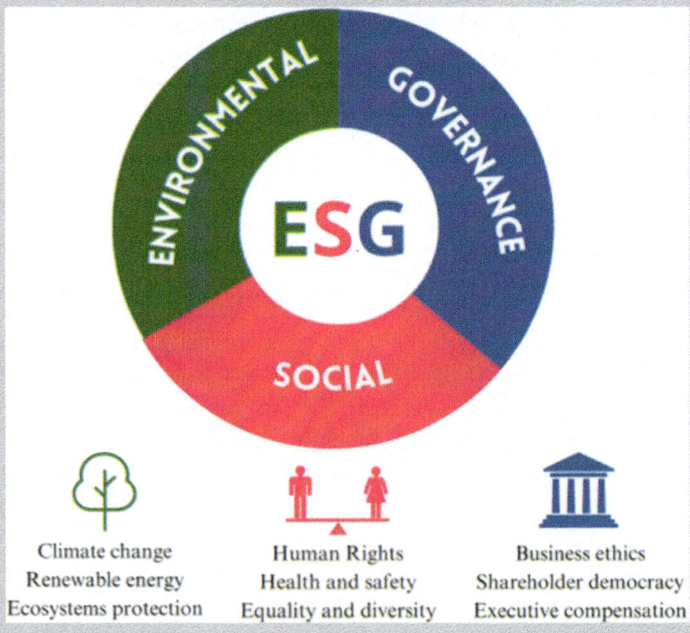

Fig. 6.1 ESG graph © Eoh-for-good

This concept is being widely used today by companies, investors and regulations. Therefore, ESG has become a crucial factor in decision-making processes across various industries. Regulations and governments are playing a significant role in promoting ESG practices. Many countries have introduced regulations and policies that require companies to disclose their ESG-related activities and performance. This transparency not only raises accountability but also allows stakeholders to make informed decisions.

Companies have recognised the importance of integrating sustainability into their operations. They understand that being environmentally responsible, addressing social issues and maintaining strong governance practices are not only ethical but also make good business sense. By adopting ESG principles, companies can enhance their reputation, attract investors and even gain a competitive advantage.

Investors have started considering ESG factors when making investment decisions, even giving rise to the figure of the Socially Responsible Investor [2]. They realise that sustainable and well-governed companies tend to be more resilient and financially sound in the long run. By incorporating ESG criteria into their investment strategies, investors can align their portfolios with their values while potentially achieving attractive financial returns.

Key recent developments include the following [3]:

In 2020, the World Economic Forum and the Big Four accounting firms released a standardised set of stakeholder capitalism metrics to make ESG reporting by companies more consistent and easier to compare.

In 2021, the European Union's Sustainable Finance Disclosure Regulation went into effect, creating new sustainability reporting requirements for financial services and investment firms.

In 2022, the United States Securities and Exchange Commission similarly proposed rules amendments with more detailed disclosure and reporting requirements for investment funds that use ESG criteria. Also, the CDSB and the SASB standards were consolidated into the International Financial Reporting Standards (IFRS) Foundation, which plans to create a unified set of IFRS Sustainability Disclosure Standards.

In 2023, the EU's Corporate Sustainability Reporting Directive went into force in January. Eventually, it will require 50,000 companies to file annual reports on their business risks and opportunities related to social and environmental issues and how their operations affect people and the environment.

By adopting a holistic and sustainable multi-level regenerative approach to innovation, initiatives could drive positive change that goes beyond mere financial returns. One of the primary outcomes of multi-level regenerative innovation is the creation of innovative products, services and processes that address pressing societal and environmental challenges. These innovations can contribute to the development of sustainable solutions that promote resource efficiency, reduce waste and pollution and mitigate the negative impacts of unfair and/or non-responsible socio-economic activities. By integrating environmental and social considerations into their innovation efforts (from a just triple transition point of view), organisations can not only drive positive change but also gain a competitive advantage in the market.

As explained in the previous chapters, the path towards shared positive outcomes and impacts necessitates several components: (1) unleashing the potential of intra-entrepreneurship; (2) driving institutional change through a collective process and (3) encouraging effective and fair deployment, outcomes and positive impacts of multi-level innovation processes of change, which requires a comprehensive and holistic approach.

6.1 The Importance of Measuring Progress and Tracking Impact Across Dimensions

This section explores the main reasons why all types of institutions, including public, private, social organisations, clusters, academia need to engage in comprehensive measurement and impact assessment across socio-economic, technological, political, environmental and governance dimensions.

If we are or want to become committed agents of change for the common good, we need to integrate the impact design, monitoring and collection from the very beginning in the everyday life of the organisation. This needs to be done at the suitable scale and with the available resources. For instance, it would be difficult for spin offs or start-ups, striving to get their business running to use many resources on monitoring and measuring in early stages. However, if the awareness of the importance of this dimension exists, little by little the design of an ad hoc methodology will start to unfold (Table 6.1).

Measuring progress and tracking impact allows institutions to **assess their performance and effectiveness for achieving their objectives**. It enables them to evaluate their strategies, initiatives and policies in relation to desired outcomes. This evaluation nurtures transparency, accountability and continuous improvement, providing a basis for informed decision-making and resource allocation.

Measurement and impact assessment **facilitate effective stakeholder engagement and collaboration**. By tracking progress and impact, institutions can demonstrate their value and contributions to stakeholders, promoting trust and support. It also enables them to identify areas for collaboration, aligning goals and interests, leveraging resources for collective impact and enhancing synergies and cooperation among diverse actors.

The enhancement of collaboration and cooperation among different participants, within organisations or innovation ecosystems, fosters, synergistic partnerships and knowledge exchange, as explained in Chap. 4. By working collaboratively, these diverse key players can pool their resources, expertise and perspectives to co-create innovative solutions that are more effective and relevant to the needs of society. This collaboration can also lead to the emergence of new business models, networks and ecosystems that support sustainable development.

Comprehensive measurement and impact tracking **generate valuable data and evidence that inform decision-making processes**. Institutions can utilise this

6.1 The Importance of Measuring Progress and Tracking Impact Across … 95

Table 6.1 Element 10 of the compass

Element of a compass		Eoh-for-good features
	10. *Scales/Rulers* Each edge of a compass may have different rulers for use with different map scales. In Australia, a compass with scales of 1:25 k and 1:50 k will work best. Use the scale that corresponds with your map's scale to determine distances	*Monitoring and evaluation system for deployment of outcomes, results and impacts*: Regular monitoring and evaluation of progress and outcomes to track the effectiveness of the strategy, identify areas for improvement and make necessary adjustments along the way Results, outcomes and impacts in processes of change within organisations and innovation ecosystems refer to the tangible and intangible changes that occur because of transformative efforts – Results indicate specific outputs, – Outcomes signify broader changes or achievements, and – Impacts denote the long-term effects and benefits generated, including social, economic and environmental dimensions

information to identify trends, strengths, weaknesses and emerging challenges across different dimensions. Evidence-based decision-making enables institutions to prioritise interventions allocating resources efficiently and addressing critical areas requiring attention, ultimately enhancing their strategic planning and policy formulation.

> It is fine to follow social and environmental governance criteria, but we see that it is not enough to align with the SDGs or invest in X sectors or in X number of initiatives. It is crucial to monitor and actively manage and measure the impact.
>
> For example: you can apply more global benchmark indicators depending on your type of investment. However, it is needed also having specific indicators for what you want to achieve (according to your theory of change, to the set agenda for the coming strategic period).

Measurement across socio-economic, technological, political, environmental and governance dimensions **contributes to sustainable development and impact assessment**. It allows institutions to assess the long-term consequences of their actions on society, the environment and novel and more coordinated governance structures. This assessment helps identify potential risks, unintended consequences and areas for improvement, promoting the integration of sustainability principles and responsible practices into institutional strategies.

Tracking progress and impact **facilitates institutional adaptation and innovation**. By monitoring performance across dimensions, institutions can identify changing trends and emerging needs and evolution of stakeholder expectations with anticipation. This information enables them to adjust strategies, develop innovative approaches and proactively respond to emerging challenges, ensuring relevance and resilience in dynamic environments.

6.2 How Can We Design Impact Follow-Up Systems and an Ad Hoc Battery of Indicators?

Organisations and innovation ecosystems can effectively assess progress, drive sustainable development and promote positive change across multiple levels and agents by establishing robust ad hoc monitoring systems.

Systemic multi-level and multi-agent processes of change within organisations and/or innovation ecosystems is a fascinating but complex task. It requires a clear understanding of the objectives and context, identification of key performance areas, selection of relevant indicators, establishment of baselines and targets, design of data collection methods, implementation of reporting and feedback mechanisms and continuous evaluation and learning.

Defining a battery of indicators and a monitoring system for different types of organisations or ecosystems require different approaches and efforts. The size, scope, maturity level, field or sector, etc., matter.

Effective design and deployment require coordinated efforts, resource allocation and the alignment of shared goals and strategies across the different actors involved in the process of change.

Although the design of an ad hoc system needs to go hand in hand with sustainability and economic results, each entity should start by questioning what is its added value for society and the planet.

First, it is crucial to strike a balance between ad hoc design and approaches that facilitate aggregation, comparison and standardisation. To avoid reinventing the wheel, it is important to invest in capacity building, leverage existing efforts and explore the support that new technologies and tested methods can offer to our specific needs. Existing indexes and methods provide a foundation upon which to develop tailored models for various initiatives, transformations and projects that we undertake. We require monitoring and dissemination mechanisms that accomplish the following objectives:

(a) demonstrate progress;
(b) incentivise and recognise exemplary practices and
(c) identify and penalise implementations that harm ecosystems and communities.

The following steps could guide the process, although it is advisable to count on experts in impact design.

Step 1. The first step in defining a battery of indicators and monitoring systems is to **clarify the objectives and the context in which the processes of change are taking place**. Decisions made at Corporations', entities, etc., positively or negatively impact the environment and society in general. Therefore, it is essential to assess each company's action holistically, i.e. individually and in its context and sphere of action. This involves understanding the desired outcomes, the scope of the change and the specific characteristics (the givens) of the organisation or innovation ecosystem. Consideration should be given to the multi-level and multi-agent nature of the processes, as well as the interdependencies and interactions between different interested parties. Key actors need to be identified and invited to lead the process as early adopters.

Step 2. Once the objectives and context are clear, it is important **to identify the key performance areas that will be monitored**. These areas should align with the goals of the processes of change and reflect the different dimensions of sustainability, such as social, economic and environmental aspects, in line with the triple transition presented in Chap. 1. For example, key performance areas could include innovation capacity, stakeholder engagement, resource efficiency, collaboration and the generation of positive social impacts.

It is important to bear in mind that 'one size does not fit all' and that every organisation, community and context will require a tailored, flexible, adaptable and feasible monitoring system. Comprehensive models require careful thought, the 'perfect system' is usually not an achievable one in terms of resources and timing.

> Define and prioritise criteria in accordance with the dimensions and the areas of action identified in the theory of change of the entity or the innovation ecosystem.
>
> Identify the requirements and regulations of the sector to guarantee compliance (directives, regulations, standards, etc.).

Step 3. Baselines and targets provide reference points for assessing progress over time in a longitudinal basis. Baselines represent the starting point, while targets set the desired level of performance to be achieved. These baselines and targets should be established for each selected indicator. Baselines can be determined through data collection, benchmarking against industry standards, best practices or expert opinions. As mentioned above, targets should be ambitious yet realistic, taking into account the organisation's or ecosystem's capacity and resources (Table 6.2).

Indicators are quantitative or qualitative measures that provide information about the performance or progress in specific areas.

It is important to **select indicators that are relevant, meaningful and aligned with the objectives and key performance areas identified earlier**. Indicators should be specific, measurable, attainable, relevant and time-bound (following the

Table 6.2 Elements 9 and 11 of the compass

Element of a compass		Eoh-for-good features
	9. *Base plate* Hard, transparent, flat surface on which the rest of the compass is mounted. It has a ruler on its edges for measuring distances on maps. Its edge is straight and is used for taking bearings and transferring them to your map	*Baseline indicators* refer to the initial set of measurable variables or parameters used to establish a reference point or starting point for monitoring and evaluating progress. These indicators serve as a benchmark to measure the current state or condition of an organisation, system, project or initiative before any interventions or changes occur. Baseline indicators provide a basis for comparison, enabling responsible actors to assess the effectiveness and impact of interventions by comparing subsequent measurements against the initial baseline data
	11. *Bearing (Index) line* Located directly above the bezel, it's also called a 'read bearing here' mark	*Target indicators*: Key Performance Indicators to plan and to track the effectiveness of the strategy, identify areas for improvement and make necessary adjustments along the way

SMART[1] formula). They should include various types of indicators used in different fields and disciplines:

1. Outcome Indicators assess the desired outcomes and changes that occur as a result of the actions taken by a programme, project, product, intervention or service.
2. Process Indicators focus on the activities, tasks or steps undertaken to achieve a specific goal. They assess the inputs, activities and outputs to monitor progress and ensure effective implementation.
3. Input Indicators measure the resources, such as financial, human or material resources, allocated to a programme, project, product, intervention or service. They provide information on the investments made and resources used.
4. Output Indicators track the immediate or direct results of a process of change, a programme or project. They quantify the tangible products, services or deliverables produced as a result of the intervention.
5. Impact Indicators assess the broader and long-term effects of a programme or intervention. They measure the overall changes or outcomes at a societal, environmental or economic level.
6. Leading Indicators are used to predict or anticipate future outcomes or changes. They provide early signals or warnings about potential trends or developments.
7. Lagging Indicators reflect past performance or changes that have already occurred. They are often used to evaluate the effectiveness or success of a programme or intervention after a certain period.

---

[1] SMART: Specific, Measurable, Attractive, Reasonable and Timed.

8. Qualitative Indicators capture non-numerical data or information. They provide insights into subjective experiences, perceptions or qualitative aspects of a phenomenon.
9. Quantitative Indicators involve numerical data or measurements. They provide objective and measurable information that can be analysed statistically.
10. It is important to note that the specific types of indicators used vary depending on the context, purpose and/or field of application.

When generating targeted KPI and suitable impact monitoring systems, key questions are:

- Have we made the right questions from the beginning?
- Are we keeping on track the defined roadmap? And are we flexible enough to adjust and include changes over time during the process?— > efficient oversight to manage and supervise operations in a competent and streamlined manner;
- How can we keep record of the progress and monitor potential impacts?
- How can we show the added value and our contribution to society and to the well-being and regeneration of the ecosystems? So how can we disseminate what we are doing and achieving?

A good analysis on these topics will help us define meaningful workable impact monitoring systems and indicators. Follow up multidimensional:

- Outcomes and results (profit, revenues, jobs created)
- Environmental, Social and Governance (ESGs) indicators and beyond, (e.g. well-being, quality of life of communities, ecosystems).
- Continuous improvement and positive changes in society towards the just triple transition.

Step 4. Data collection methods should be designed to gather relevant information for each indicator. This can include surveys, interviews, focus groups, observation and the collection of existing data from various sources. It is important to ensure that data collection methods are reliable, consistent and suitable for the specific context (e.g. definition of a data management plan, data sets, sources of information, personal data, etc.). Careful though should also be given to data privacy and ethical considerations.

Step 5. **Develop reporting and feedback mechanisms** which are essential for communicating progress, sharing information and promoting accountability. In this sense, it is desirable for each organisation to communicate its mission, vision and values and provide objective and transparent information about its activity. What is not shown, it does not exist and its return on investment cannot be recalled.

Regular reports should be produced to provide updates on the performance of the selected indicators. These reports can be shared internally within the organisation or ecosystem and externally with relevant actors, such as investors, policy makers and the public. Feedback mechanisms should be established to gather input, insights and suggestions for improvement from interested parties.

Step 6. The monitoring system should be implemented to track the selected indicators over time. Effective monitoring, enforcement and accountability mechanisms are necessary to ensure compliance and address any potential negative impacts.

Continuous evaluation and learning are essential for improving the battery of indicators and monitoring systems. This involves the regular collection, analysis and interpretation of data. What should be done?

1. Regular evaluations conducted to assess the effectiveness, relevance and impact of the indicators and monitoring processes. Monitoring should be conducted at appropriate intervals, allowing for timely identification of trends, drivers, challenges and opportunities. The monitoring system should be flexible and adaptable, allowing for adjustments and refinements based on changing circumstances or new information.
2. Lessons learned captured and adjustments made to improve the system.
3. Stakeholder feedback and engagement that can provide valuable insights for evaluating and refining the monitoring system.

Encouraging effective and fair deployment, outcomes and positive impacts of multi-level regenerative innovation processes requires a combination of collaboration, stakeholder engagement, robust regulatory frameworks, equitable access to resources, continuous evaluation and knowledge sharing. Actors can work together to ensure that regenerative innovations are implemented in a way that addresses social inequalities, respects ethical considerations and maximises sustainable development outcomes. Such efforts are essential for creating a just and inclusive future where the benefits of innovation are shared by all.

References

1. Benczur P, Kvedaras V, Preziosi NMC (2023) Health-adjusted income: complementing GDP to reflect the valuation of life expectancy. Publications Office of the European Union, Luxembourg. https://doi.org/10.2760/91140
2. Mansour K (29 Oct, 2020) ESG ratings: how can a business' environmental and social impact be measured? Early Metrics. https://earlymetrics.com/esg-ratings-how-can-a-business-environmental-and-social-impact-be-measured/. Accessed 2 July 2023
3. Mathis S, Stedman C (n.d.) What is environmental, social and governance (ESG)? WhatIs.com. https://www.techtarget.com/whatis/definition/environmental-social-and-governance-ESG. Accessed 27 June 2023

Open Access This chapter is licensed under the terms of the Creative Commons Attribution 4.0 International License (http://creativecommons.org/licenses/by/4.0/), which permits use, sharing, adaptation, distribution and reproduction in any medium or format, as long as you give appropriate credit to the original author(s) and the source, provide a link to the Creative Commons license and indicate if changes were made.

The images or other third party material in this chapter are included in the chapter's Creative Commons license, unless indicated otherwise in a credit line to the material. If material is not included in the chapter's Creative Commons license and your intended use is not permitted by statutory regulation or exceeds the permitted use, you will need to obtain permission directly from the copyright holder.

Chapter 7
Conclusion: Organisations and Ecosystems in Transition—Nurturing Transformative Governance

In this transformative journey, we have explored the crucial aspects of managing change and navigating the complexities of emerging systems and innovations. Throughout the chapters, three key insights have emerged, guiding us towards a more just, sustainable and adaptive future.

First and foremost, we have recognised the urgency to plan with anticipation and tackle profound transitions head-on. We cannot wait until it is too late, until our organisations are in the declining phase, our ecosystems in dangerous turning points and our innovations in the valley of death. The era of transitions we are facing demands proactive action and foresight to address the challenges and opportunities that lie ahead. By acknowledging the transition gap between established and emerging systems, we gain a deeper understanding of the disparities and complexities that need to be addressed. The multidimensional co-creation vortices of transition have illuminated various facets of transformative change, highlighting the need for holistic and inclusive approaches.

> The processes of change require professional mediation by experts such as tecno-anthropologists and organisations like Eoh-for-Good. These specialists possess the unique ability to bridge the gap between different disciplines, sectors, innovators and types of organisation and actor, understanding the intricate interactions between humans and natural and innovation ecosystems. Their expertise facilitates a deeper comprehension of human needs and behaviours, ensuring that the changes implemented are not only technologically sound or profit driven but also socially and culturally sensitive. Eoh-for-Good's commitment to positive impact aligns with the goals of fostering a just triple transition, being an ideal partner to navigate complex transitions with more conscious, ethical and sustainable solutions.

Secondly, the concept of a Just Triple Transition has emerged as a guiding principle for transformative governance. By balancing social, green and digital considerations, we can create a harmonious and integrated approach to address the interconnected challenges of the future. The compass of transformative governance, grounded in approaches and principles for the common good, can provide us with a navigational tool to steer our joint efforts in the right direction.

These governance principles and ways of doing need to be tailored to the contexts and the specific needs. They will help address the urgent hilly or steep path to transformation being concreted in numerous initiatives or within strategic plans:

1) Resilience of critical infrastructures (renewable energies, preparedness to cyber-attacks, etc.,
2) Health and well-being (detention, prevention and involvement of citizens into their health care and
3) Education, capacity building in the lifelong learning (up-skilling, reskilling, nurturing, entrepreneurship from the childhood)

Cultivating a culture of innovation and risk-taking through intra-entrepreneurship entrusts individuals and organisations to become agents of transformative change. Leadership plays a pivotal role in shaping an inclusive and adaptive organisational culture, raising an environment where innovative solutions can thrive, giving voice and agency to all parties. Diversity and co-creation thrive innovation.

> In an era of profound transitions and rapid change, multi-i dynamic co-creative capabilities and dynamics are becoming vital for organisations to effectively navigate evolving environments. Organisations will be better equipped to adeptly respond to the challenges posed by the triple transition by integrating, building and reconfiguring internal and external competences and entrepreneurial capabilities. Dynamic capabilities enable agility, adaptability and resilience, ensuring that companies and organisations of different nature can seize opportunities and effectively address emerging complexities. This can maximise the potential for success and sustainable just growth.

Lastly, collaborative elements that start with an 'i' have emerged as the driving forces of systemic transformative governance. Co-creative vortices of innovation, formed through collaboration between diverse stakeholders, are hubs of transformative solutions. Furthermore, navigating the innovation processes through learning feedback and feed-forward loops and multi-level and multi-agent governance processes become more impactful. Designing more targeted and ad hoc impact monitoring systems will allows us to measure progress and track the outcomes of transformative initiatives, ensuring evidence-based decision-making, proving return on investment and maximising the power of entrepreneurship and innovation.

As European society, we can lead the way and serve as a huge 'collaboratory' for the all-encompassing just triple transition. This would require:

1. **Collaboration across sectors, disciplines and communities to drive multi-stakeholder engagement, collective intelligence, harness diverse perspectives and leverage the full spectrum of knowledge and expertise.** By involving a wide range of actors (from all ages, different professions, diverse fields and sectors, backgrounds and contexts, etc.), we can co-create a) shared and negotiated visions and b) comprehensive and inclusive solutions that address complex challenges.
Establishing stronger links with other companies, labs and entities reduces the chances of failure of our Minimum Viable Products or Services. This is more critical in the case of innovations with a clear social impact, as the failure can have more negative effects than in the case of purely commercial innovation. This is also related to the paradoxical thought of testing innovations in an increasingly agile way while minimising uncertainty using methodologies born in the business context and that are already being used for the development of social innovation. We need to generate novel innovations to address the triple impact, economic, social and environmental in a more comprehensive, sustainable and inclusive way.
2. **Unfolding and experimenting with the quadruple (or n-) helix model that expands the triple helix model of innovation by adding the dimension of civil society as a key factor in the innovation ecosystem.** This model recognises the importance of collaboration and co-creation among academia, industry, government and civil society to drive innovation and societal progress. By actively involving civil society organisations, community groups and citizens in the innovation process, actors committed to change can foster inclusivity, diversity and collective ownership of innovation outcomes.

There are still many challenges to be solved and solutions for how all these governance issues are addressed. The idea is to leverage the collective expertise and collaboration across academia, industry, government and civil society to drive transformative change. Work towards building sustainable ecosystems fostering environmental regeneration with the goal of preserving and revitalising the environment for long-term ecological balance and well-being.

As we conclude this journey, we envision European society as a huge collaboratory—a space where multi-i governance fosters positive change on a continental scale. The transformative governance framework presented in this book offers a pathway for shaping a more resilient and adaptive future.

In conclusion, I hope that the insights gained from this book can entrust us to become agents of change fostering resilient and adaptive governance for the future. By applying the principles of transformative governance, we can navigate the complexities of profound transitions and shape a more just, sustainable and inclusive world for generations to come. Referencing the illustrative experiences and concrete cases presented in this book, we can enact innovation in our organisations and create robust liaisons to make innovation consciously sustainable. As we continue this transformative journey, let us embrace the spirit of collaboration and innovation, working together towards a brighter and better future for humanity and the planet.

> Discovering what inspires us, driven by our unique talents, abilities, interests, knowledge, motivation and aspirations, is essential. By aligning our passions with those of others, we can engage in an everlasting dance of cooperation and competition, working together to make our world a better place.

Open Access This chapter is licensed under the terms of the Creative Commons Attribution 4.0 International License (http://creativecommons.org/licenses/by/4.0/), which permits use, sharing, adaptation, distribution and reproduction in any medium or format, as long as you give appropriate credit to the original author(s) and the source, provide a link to the Creative Commons license and indicate if changes were made.

The images or other third party material in this chapter are included in the chapter's Creative Commons license, unless indicated otherwise in a credit line to the material. If material is not included in the chapter's Creative Commons license and your intended use is not permitted by statutory regulation or exceeds the permitted use, you will need to obtain permission directly from the copyright holder.

Bibliography

1. AICBR. (n.d.). AICBR. https://www.aicbr.ca. Accessed 19 June 2023
2. Bianchi G, Pisiotis U, Cabrera M (2022) GreenComp, the European sustainability competence framework. Publications Office of the European Union, Luxembourg. https://data.europa.eu/doi/10.2760/13286. Accessed 14 July 2023
3. Bizkaia 2050: una visión de futuro del Territorio (28, April 2022) Cámarabilbao | Cámara de Comercio de Bilbao. https://www.camarabilbao.com/corporativo/bizkaia-2050-vision-futuro-territorio-202204281232/. Accessed 28 Nov 2022
4. Blicharz GJ (2017) Decodification, common good, and responsible societies: beyond the Elinor Ostrom's theory of governing the commons. In: Vargas AI, Alonso-Bastarreche G, Van Schalkwijk D (eds) Transcendence and love for a new global society. Servcio de Publicaciones de la Universidad de Navarra 2017, Pamplona, pp 91–101. https://www.academia.edu/43408738/Decodification_common_good_and_responsible_societies_beyond_the_Elinor_Ostrom_s_theory_of_governing_the_commons. Accessed 19 June 2023
5. Cann O (2020) Measuring stakeholder capitalism: world's largest companies support developing core set of universal ESG disclosures. World economic forum. https://www.weforum.org/press/2020/01/measuring-stakeholder-capitalism-world-s-largest-companies-support-developing-core-set-of-universal-esg-disclosures/. Accessed 11 July 2023
6. Col·laboratori Catalunya (n.d.) Col·laboratori Catalunya. https://colabscatalunya.cat/. Accessed 19 June 2023
7. Covey SR (1999) The 7 habits of highly effective people. Touchstone, United Kingdom
8. Davila T, Epstein MJ (2014) The innovation paradox why good businesses kill breakthroughs and how they can change. Business Book Summarie. Accessed 3 July 2023
9. European Commission (n.d.) Inforegio—Interreg C—interregional cooperation. Interreg C—Interregional cooperation. https://ec.europa.eu/regional_policy/policy/cooperation/european-territorial/interregional_en. Accessed 20 June 2023
10. European Investment Bank (2023) Investment report 2022/2023: resilience and renewal in Europe. https://www.eib.org/attachments/lucalli/20220211_economic_investment_report_2022_2023_en.pdf. Accessed 1 July 2023
11. Foran MP (21 Oct, 2022) Equal dignity and the common good. SSRN Scholarly Paper, Rochester, NY. https://doi.org/10.2139/ssrn.4254952
12. Global Gateway (1 March, 2023) https://commission.europa.eu/strategy-and-policy/priorities-2019-2024/stronger-europe-world/global-gateway_es. Accessed 14 July 2023
13. Habilidades interpersonales y selección de personal | EAE (16 April, 2020) Retos en supply chain | Blog sobre supply chain de EAE business school Barcelona. https://retos-operaciones-logistica.eae.es/las-habilidades-interpersonales-y-la-seleccion-de-personal/. Accessed 19 June 2023
14. Hobbes T (2020) Leviathan

15. https://www.facebook.com/trailhiking. (5 March, 2019) Anatomy of a hiking compass | Hike navigation. https://www.trailhiking.com.au/navigation/anatomy-of-a-compass/. Accessed 11 July 2023
16. https://www.innerdevelopmentgoals.org. Accessed 19 June 2023
17. INTEGER (2023) https://integercollab.eu/. Accessed 13 June 2023
18. Intercultural Collaborations (n.d.) Intercultural collaborations. https://interculturalcollaborations.com. Accessed 20 June 2023
19. Intergovernmental Panel on Climate Change (2023) IPCC—intergovernmental panel on climate change. https://www.ipcc.ch/. Accessed 14 July 2023
20. Interngovernmental Panel on Climate Change (IPCC) (2023) Climate change 2023 synthesis report (A report of the intergovernmental panel on climate change). Geneva, Switzerland, p 36. https://www.ipcc.ch/report/ar6/syr/downloads/report/IPCC_AR6_SYR_SPM.pdf. Accessed 1 July 2023
21. Interngovernmental Panel on Climate Change (IPCC) (n.d.) Climate change: a threat to human wellbeing and health of the planet. Taking action now can secure our future—IPCC. https://www.ipcc.ch/2022/02/28/pr-wgii-ar6/. Accessed 14 July 2023
22. Intersectoral collaboration (n.d.) Novartis foundation. https://www.novartisfoundation.org/urban-population-health-toolkit/intersectoral-collaboration. Accessed 20 June 2023
23. Locke J (1823) Two treatises of government, vol V. McMaster University Archive of the History of EconomicThought. https://www.yorku.ca/comninel/courses/3025pdf/Locke.pdf
24. Mulligan S (2010) Capabilities and the common good. Ir Theol Q 75(4):388–406. https://doi.org/10.1177/0021140010377739
25. Osborne T (2008) MacIntyre, thomism and the contemporary common good. Analyse Kritik 30(1):75–90. https://doi.org/10.1515/auk-2008-0105
26. Pagoropoulos A, Pigosso DCA, McAloone TC (2017) The emergent role of digital technologies in the circular economy: a review. Procedia CIRP 64:19–24. https://doi.org/10.1016/j.procir.2017.02.047
27. Pan-European Matchathon (n.d.) https://euvsvirus.org. Accessed 11 June 2023
28. Petrevska R, Caro-Gonzalez A, Bertello A, Bogers M (2022) Academia diffusion experiment: trailblazing the emergence from co-creation: open innovation and open collaboration towards ecosystem building. In: Facilitation in complexity: from creation to co-creation, from dreaming to co-dreaming, from evolution to co-evolution. Springer International Publishing. https://www.researchgate.net/publication/366979383_Academia_Diffusion_Experiment_Trailblazing_the_Emergence_from_Co-Creation_Open_Innovation_and_Open_Collaboration_Towards_Ecosystem_Building. Accessed 11 May 2023
29. Rawls J (1971) A theory of justice. Harvard Univ Press. https://doi.org/10.2307/j.ctvjf9z6v
30. Sandel MJ (2006) Public philosophy: essays on morality in politics. Harvard University Press, Cambridge, Mass
31. Sayre NF (2012) The politics of the anthropogenic. Annu Rev Anthropol 41(1):57–70. https://doi.org/10.1146/annurev-anthro-092611-145846
32. Senge PM (2006) The fifth discipline: the art and practice of the learning organization (2nd edn). Random House Business
33. United Nations Department of Economic and Social Afairs (n.d.) The 17 goals | sustainable development. https://sdgs.un.org/goals. Accessed 28 Oct 2022
34. United Nations General Assembly U (1948) Universal declaration of human rights. United Nations. United Nations. https://www.un.org/en/about-us/universal-declaration-of-human-rights. Accessed 20 June 2023
35. URBANOME (n.d.) URBANOME: urban health, wellbeing, liveability. https://www.urbanome.eu/. Accessed 23 June 2023
36. Warr M, West RE (2020) Bridging academic disciplines with interdisciplinary project-based learning: challenges and opportunities. Interdisc J Prob-Based Learn 14(1). https://doi.org/10.14434/ijpbl.v14i1.28590
37. Wilson (10 June, 2023) Why is collaboration so important in the building of scientific knowledge. https://collaboratory.ist/why-is-collaboration-so-important-in-the-building-of-scientific-knowledge/. Accessed 1 June 2023

38. Wulf WA (1989) The national collaborator—a white paper. In: Lederberg J, Uncaphar K (eds) Towards a national collaboratory: report of an invitational workshop at the Rockefeller university, march 17–18 (appendix A) national science foundation, directorate for computer and information science engineering, Washington, D.C. Accessed 13 June 2023

The manufacturer's authorised representative in the EU is Springer Nature Customer Service Centre GmbH, Europaplatz 3, 69115 Heidelberg, Germany. If you have any concerns regarding our products, please contact ProductSafety@springernature.com

Printed and bound by CPI Group (UK) Ltd, Croydon, CR0 4YY
23/03/2026
02076369-0004